Civic Engagement in Asia

Mochamad Indrawan · Jeffrey B. Luzar ·
Helen Hanna · Theodore Mayer
Editors

Civic Engagement in Asia

Transformative Learning for a Sustainable Future

Buku OBOR

Springer

Editors
Mochamad Indrawan
Research Center for Climate Change
Universitas Indonesia
Depok, Indonesia

Helen Hanna
Manchester Institute of Education
University of Manchester
Manchester, UK

Jeffrey B. Luzar
Independent Human Ecologist
Lizton, IN, USA

Theodore Mayer
Institute for Transformative Learning of the International Network of Engaged Buddhists, Thailand

Supported by Chulalongkorn University, Universitas Indonesia's Research Center for Climate Change (RCCC—UI), the Asia-Pacific Network for Global Change Research (APN), and Samdhana Institute.

ISBN 978-981-16-9383-0 ISBN 978-981-16-9384-7 (eBook)
https://doi.org/10.1007/978-981-16-9384-7

Originally published by Yayasan Pustaka Obor Indonesia

This Springer imprint is published by the registered company Springer Nature Singapore Pte Ltd.
The registered company address is: 152 Beach Road, #21-01/04 Gateway East, Singapore 189721, Singapore

Foreword by Surichai Wun'gaeo

As we enter the third decade of the twenty-first century, there is a widespread sense that we live in unprecedented times, a turning point of some kind. The COVID-19 pandemic has dramatically heightened this sense and made it more palpable everywhere. The pandemic has also highlighted the pressing need for genuine solutions to the mounting problems we face together now as human beings. Yet, in facing these complex challenges, there is at the same time a strong sense of disconnectedness, not only in national but also in regional and international contexts. In other words, we are in urgent need of increased solidarity to envision our common future.

If you are a decision-maker, the words I have to say here, and indeed this entire volume, are addressed in large part to you. Yet they are addressed to you primarily as a fellow human being and fellow citizen who has chosen to take responsibility, within your capacity, for our common plight. In this sense, it will be of benefit to all who, like you, wish to learn in each other's company *how* we can best take responsibility together, and *how* we can thus extricate ourselves from our shared dilemmas.

Holding the reins of power, especially centralized power, one faces two substantial obstacles in working for progress or solving societal problems. The first is that one cannot know, much less feel as a fellow human being, all the myriad difficulties, sufferings, and challenges that are experienced by the inhabitants of the many communities arrayed within even a single province or country, much less an entire region or the planet as a whole. As a result, one may not be moved to respond or have no clear grounds for constructing a logical response. The second obstacle consists of the coercive power structures in our societies that prohibit more engaged, thoughtful, or innovative contributions by the people on the ground to the solution of complex social problems. Unfortunately, these structures have had a steady impact on people, creating a deficit of trust. My question is: Without trust, how can we create or even imagine a common future?

There is another kind of power in the social arena that can help us overcome these obstacles. It is the capacity to be inspired by what others have dared to think and to do, and to *learn* from them. It is a transformative power that allows individuals, communities, and entire societies to learn together. This is one of the key meanings of the term "transformative learning" in my view, and it is at the heart of civic engagement. Open and unfettered communication is essential for it to work. It also requires that each individual, community, and form of life even—taken within its own particular context of aspirations and challenges—is accorded the dignity and respect appropriate to living beings in their own right. Yet each then also becomes potentially a source of information, companionship, knowledge, support, solutions, and a sense of what is possible. This understanding of civic engagement as a kind of societal learning based on mutual respect allows us to pursue the creation of a sustainable and just society with dignity for all by enacting it.

The concept of transformative learning is indebted to Jack Mezirow, who is responsible for a historic shift in adult education—from focusing on skills development to cultivating profound transformations in individuals' sense of themselves, of their society, and of what they could do. For Mezirow, transformative learning was often brought on by an existential dilemma, such as the death of a loved one or the loss of a job; but it could also be inspired by an "eye-opening" discussion, painting, or poem, or by encounters with cultural assumptions that challenged one's own (Mezirow 1991, 1998). All such transformations involved a profound questioning of deeply-held assumptions and perspectives. To come to fruition, the learning entailed by such shifts in perspective also required open exchanges among peers, guided by the search for meaning and for genuine solutions (rather than by the imperatives of the market or the state),[1] and aiming toward action.

This is precisely what the reader will find in this volume: reflections and lessons shared by those who have identified a social or ecological dilemma in Asia. The volume is a collection of original voices from the ground, through the keen eyes of experienced professionals, many of whom are founders of NGOs or social enterprises, and some of whom have worked within government for policy reform. All speak as insightful witnesses to the challenges and hopes of civic engagement within a rapidly changing Asia.

As a participant over the years in multidisciplinary platform exchanges and social movements in the region, I heartily welcome the shift this volume represents—a sharing of diverse experiences as we face pressing common issues and learn from our engagement with them. While all the authors bring academic rigor to their writing, their focus is not academic theorizing per se, but rather the need for joining together to better understand and alleviate the many forms of suffering in our region.

It was concerns on the part of academic communities within Asia that set the stage for this volume, concerns that arose as we took stock of the many crises facing us. These include emerging diseases, global warming and biodiversity loss, toxic waste, unsustainable urban development, income and wealth inequality, economies focused on extraction and growth at all costs, and poor quality education or lack of access to it entirely. The need to address the common issues that our region faces called for the kind of collective learning process that is often realized most fully in contexts of civic engagement.

[1] Freedom from market and state imperatives is present at least implicitly in Mezirow, but explicit and accentuated in Habermas (Fleming 2000).

Our journey began in October 2015, when Michiko Yoshida and I invited a small group of civil society leaders and academics from Malaysia, Thailand, Indonesia, and Japan to gather at Chulalongkorn University in Thailand. We agreed upon a common goal, namely, to promote strategic and participatory processes that could foment transformative learning to achieve a just and ecologically sustainable community in the region. The transformations would have to include not only changed ways of thinking and behaving, but also systemic reform in public policy and institutions. Policy advocacy in turn needed to be based upon both adequate evidence and sustainable planning. We sought to encourage broader people's participation in constructing our community of practice. As such, we wanted to facilitate the work of public intellectuals who were actively engaged in regional struggles and social movements. We envisioned multi-sectoral collaborations of key players among academics and practitioners at various levels. Making a solid case for collaborative multi-sectoral action also needed a carefully designed series of interventions, and these constitute the core of our current approach.

Concretely, we sought to achieve these goals through a series of dialogues leading to a regional workshop entitled "*Civic Engagement for a Just and Sustainable ASEAN: Our Stories and Practices*" (Yogyakarta, August 11–15, 2017). Jointly organized by the Institute of Asian Studies at Chulalongkorn University and the Indonesian Consortium for Religious Studies, this undertaking brought together grassroots innovators and social entrepreneurs, as well as researchers and policy professionals. Ably facilitated by Toshiyuki Doi (Mekong Watch) and Yeoh Seng Guan (Monash University), the workshop gave rise to energetic exchanges as well as follow-up collaborations between participants. This workshop dialogue then became the basis for the development of this edited volume. Aiming to enrich the perspectives of the original workshop participants, Mochamad Indrawan, the editor, subsequently enlarged the range of contributions to the volume.

While this book is one outcome of the Yogyakarta workshop, that same gathering became the impetus for further frameworks for collaboration, including the Bangkok Forum and Civic Engagement 4.0. The Bangkok Forum was launched in 2018 by Chulalongkorn University with the support of the Korea Foundation for Advanced Studies (KFAS) as an integrative knowledge platform with the theme of "Future Sustainable Asia." The Forum brought civil society actors into meaningful engagement with policymakers and university professionals, with special attention to the social dimensions of sustainability. More than 800 participants joined the conference, and special efforts were made to include young adults. The Civic Engagement 4.0 platform, launched as the International Forum in Solo, Indonesia, August 21–22, 2019, brought together some 380 participants. This gathering combined academic panels with hands-on workshops and a mayors' symposium to reflect on sustainability issues from the practical perspectives of governance. The Forum relied integrally on the planning and coordination of Kota Kita Foundation, a Solo-based Indonesian CSO working for citizen empowerment and participatory urban planning (Chapter 10 of this volume). Kota Kita's dynamic engagements in Solo attracted large numbers of enthusiastic students and other young adults as volunteers and participants.

The production of this book, therefore, has been one part of a much larger process of thought, action, and relationship building on the part of many people who have focused their efforts on transformative learning and civic engagement in Asia. This book is the testimony of those committed to creating solidarity as their ethical responsibility for the future. I invite you now to join the larger process of change and reflection of which this volume is a small but significant part, by stepping into its pages, guided by your own interests and curiosity.

Surichai Wun'gaeo
Professor Emeritus of Sociology
Director of the Peace and Conflict Studies Center
Chulalongkorn University, Bangkok, Thailand

References

Fleming, T. 2000. "Habermas, Democracy and Civil society: Unearthing the Social in Transformative Learning". In *Challenges of Practice: Transformative Learning in Action*, edited by C. Wiessner, S. Meyer, and D. Fuller, 303–308. New York: Columbia University.

Mezirow, J. 1991. *Transformative Dimensions of Adult Learning*. San Francisco: Jossey Bass Publishers.

Mezirow, J. 1998. "Transformative Learning and Social Action: A Response to Inglis." *Adult Education Quarterly* 49, no. 1: 70–73.

Foreword by Jatna Supriatna

The world is shaped not by individuals alone, but by networks of stakeholders including government, business, civil society, and academia. As diverse stakeholders, we shape the world we share through individual actions such as the foods we eat and the energy we use. We also shape our world through our collective interactions with a web of institutions that interconnect the world.

A sustainable world can be achieved if we begin thinking differently. This means we are no longer to think of ourselves as merely autonomous actors, but also as significant nodes within a larger institutional web. The decisions we make, the clothes we wear, the community efforts we join are all mediated through that intricate web. The world is interconnected and borderless. There is no single country, organization, individual, or machine that can stand alone or change the world. We need to collaborate in order to close the gap between the many emerging issues and our own understanding and response to them.

Society needs to stabilize and govern the dynamic challenges of sustainability. This book inspires us by demonstrating what individuals and communities can do with the might of the mind and spirit so as to encourage transformative learning that can then intensify civic engagement. The stories collected here relate rich and diverse examples of transformative learning, for instance:

- A Thai professor who went out of her comfort zone and for more than ten years undertook field research and outreach to increase beach-goers' fighting chance against the deadly box jellyfish.
- An activist who went to great effort to support artisans from the Indonesian island of Sumba to produce world class woven clothing, while finding means to better connect the industry with nature, indigenous culture, and sustainability.
- On-the-ground activists from Fukushima who taught concerned citizens how to increase self-reliance in the face of a deadly triple disaster the modern world has not seen before.
- A Buddhist scholar whose compassion for people helped Asian young adults transform themselves and revolutionize their way of thinking.

These are just a few inspiring examples of how Asia's public intellectuals have readily lent themselves to transformative learning and civic engagement in order to create a just and socially equitable future for Asia.

These 24 chapters, each written by true fighters and public intellectuals, support the maxim that the whole is greater than the sum of its parts. The expert contributors and the international editors—Mochamad Indrawan (Indonesia), Jeffrey Luzar (USA), Helen Hanna (UK), and Theodore Mayer (Thailand/ USA)—are to be congratulated for putting their hearts and minds together to produce such a vivid and colorful perspective.

Jatna Supriatna
Founder and Chairperson
Institute of Sustainable Earth and Resources
Universitas Indonesia, Depok, Indonesia

Foreword by Julian Caldecott

This book is a rich source of the thoughtful case studies from which interdisciplinary understanding and transformative learning are built. It draws on the life- and work-experiences of expert observers in diverse social and geographical circumstances across Asia. In the process, it makes original contributions to thinking about relationships between people and nature, and between people. The whole volume takes a welcome syncretic approach to ecology, spirit and community, reminding us that people live in worlds created through their own culture in dialogue with mystery, history and ecology. In exploring this from diverse points of view, the authors give attention to key tasks and challenges, including in the areas of advocacy, governance, citizen science, tradition, faith, leadership, and education. They base their thinking on many years' engagement with specific peoples, often alongside deeply experienced civil society organisations. By exploring the essential issue of 'meaning', they constructively fuse the two key themes of ecology and anthropology. The approach gives rise, for example, to the useful notion of 'heartware' in Chapter 3, which represents the emotional domain surrounding reason, wisdom and felt experience in a lived environment.[2]

The community orientation is consistent and strong, reminding us that our lives really only make sense in a social and ecological context. The result is that the book bears comparison with the foundational volume *The Wealth of Communities: Stories of Success in Local Environmental Management* by C. Pye-Smith and G. Borrini-Feyerabend (1994). This places it within a current renaissance of appreciation for community-based environmental management, which is fast becoming prominent as a key way for societies to adapt to climate change and ecological chaos. This theme, for example, has grown rapidly within government adaptation plans sent to the UNFCCC since the 2015 Paris Agreement. They include submissions from Costa Rica, Dominica, Fiji, Ghana, Guinea-Bissau, Kiribati, Lesotho, Namibia, Nepal,

---

[2] Ecologist, environmental sustainability consultant, Director of Creatura Ltd, Fellow of the Schumacher Institute for Sustainable Systems, and author of *Designing Conservation Projects* (2009), *Aid Performance and Climate Change* (2017), *Water: Life in Every Drop* (2020), and *Surviving Climate Chaos by Strengthening Communities and Ecosystems* (2021).

Nigeria, Suriname, South Sudan, and Tonga. But the fact that apart from Nepal none of these stand-out examples are in Asia suggests the need for Asian stakeholders to arm Asian and ASEAN governments with a better understanding of these issues and opportunities, and to press them further and harder. This is an urgent task to which *Civic Engagement in Asia* is well able to contribute, responding to the need for all communities and nations to survive climate chaos inclusively, in their own circumstances and on their own terms.

For it is worth remembering that the topics addressed in this book, and the examples offered and explored, exist in a world that is now under threat as never before. The climate and ecological emergency, and world-wide public responses to it, are now critical and will remain so for the rest of our lives and those of our children. Our responsibility now is to secure the future and to build a sustainable world, without sacrificing the values of diverse human cultures that make life worth living. This can only be done though respectful understanding of diversity, and recognition that solutions can come from many sources, informed by inclusive engagement and traditional knowledge. These are facts that any reader of this book will learn to appreciate at depth. As a knowledge resource, therefore, it will be useful to anyone working in, teaching about, or seeking careers in the burgeoning applied field of sustainable development and climate change. With luck, many of its ideas will find their way into the minds of decision makers who are now shaping the future of all the Asian environments and peoples.

April 2021

Julian Caldecott
Director of Creatura Ltd and
Fellow of the Schumacher Institute for
Sustainable Systems, Edinburgh, Scotland

Reference

Pye-Smith, C., and G. Borrini-Feyerabend. 1994. *The Wealth of Communities: Stories of Success in Local Environmental Management*. Abingdon: Routledge.

Acknowledgments

This volume represents a compilation of 24 articles by 31 authors from Asian countries. The editors would like to thank the authors for their commitment and patience throughout the lengthy review and revision process.

Insightful input from Surichai Wun'gaeo, Michiko Yoshida, Yeoh Seng Guan, and Toshiyuki Doi helped develop the primary focus for *Civic Engagement in Asia*.

Fifteen of the 24 volume chapters were sourced from the workshop "*Civic Engagement for a Just and Sustainable ASEAN: Our Stories and Practices*" (Yogyakarta, August 11–15, 2017), which was jointly organized by Chulalongkorn University (Thailand) and the Indonesian Consortium for Religious Studies. The sessions were ably facilitated by Toshiyuki Doi and Yeoh Seng Guan, with further support from Dicky Sofjan (ICRS and Universitas Gadjah Mada). Toshiyuki Doi collated the primary materials from the workshop.

Editorial work was provided by an international team led by Mochamad Indrawan. Technical editing was provided by Jeffrey Luzar, Helen Hanna, and Theodore Mayer. Copy editing was conducted by Jackie Imamura, Stephen Lee Mowat, Melissa Storms, Dexter Bohn, and Andre Haryono. Japanese language research assistance was kindly offered by Saki Fujioka.

Expert reviews of the chapters were provided by Toshiyuki Doi, Jeremy Ironside, Agus Sumule, and Nadarajah Manickam.

Support for the management of the project was provided by Budi Haryanto and Nurul Winarni, along with Maya Dewi from the Research Center for Climate Change—Universitas Indonesia. We are likewise very grateful for the consistent support given by Nualnoi Treerat, Director of the Institute of Asian Studies at Chulalongkorn University, and Pirongrong Ramasoota, Vice President for Social Outreach and Global Engagement of Chulalongkorn University during the development of this volume.

We also thank Kartini Nurdin and the team from Yayasan Pustaka Obor for their professional support and prompt processing of the manuscripts.

Thanks are also due to Record Indonesia (Rekam Nusantara) and Arbain Rambey for the photographs they contributed to this volume. Andi Harris Firmansyah led the development of the cover design. Neni Rochaeni and Andhika Vega Praputra from Samdhana Institute provided invaluable support for the overall design of the book.

The development of this edited volume received generous support from the Asia Pacific Network for Global Change Research (APN), from Chulalongkorn University, and from Universitas Indonesia's Research Center for Climate Change (RCCC—UI). Samdhana Institute further facilitated the production of this book. Responsibility for the views presented in this volume, along with any mistakes or failings, lies of course with the relevant authors and editors themselves.

Finally, we wish to appreciate the encouraging words from Linda Stevenson of APN that "APN is delighted to endorse this publication as an outcome of a Capacity Development Programme (CAPaBLE)-supported project, CBA2018-05SY-Indrawan."

Contents

Chapter 1
Building Livelihood Sovereignty for the Mekong Region

Tran Thi Lanh

Highlights

- Indigenous ethnic minorities in the Mekong region are struggling to maintain their local values and rights for livelihood sovereignty.
- The sustainability values of the indigenous peoples, which have been degraded in many parts of the world, need to be preserved and supported.
- Government support for legalizing the voluntary, community, and customary law-based administration of natural resources by the local people needs to be rallied through intricate advocacy and facilitation.
- The key to building the rights-based approach was developing strong networks of "key farmers" to act as effective speakers in farmer-to-farmer and farmer-to-authority workshops.
- A significant achievement was the passing of Vietnam's 2017 Law on Forestry, which now stands as a springboard for a new series of actions by NGOs on behalf of indigenous ethnic minority communities in Vietnam.

Introduction

The Livelihood Sovereignty Alliance (LISO) is an alliance of three Vietnam-based civil society organizations: the Social Policy Ecology Research Institute (SPERI), the Community Entrepreneur Development Institute (CENDI), and Culture Identity and Resource Use Management (CIRUM). Each of these is dedicated to working toward the livelihood sovereignty of indigenous ethnic minorities in the Mekong region (see Appendix A).

T. T. Lanh (✉)
Social Policy Ecology Research Institute (SPERI), Hanoi, Vietnam
e-mail: ttlanh@speri.org

M. Indrawan et al. (eds.), *Civic Engagement in Asia*,
https://doi.org/10.1007/978-981-16-9384-7_1

"Livelihood Sovereignty" is defined[1] here as the right of people to their land, religion, culture, knowledge, and system of governance. Livelihood sovereignty can be considered "a holistic and ethical alternative solution," which consists of five interrelated rights: (1) the right to land, forest, water, clean air, and natural landscape, (2) the right to maintain one's religion, (3) the right to live according to one's own way of life and values of happiness and well-being within one's own natural environment, (4) the right to operate according to one's own knowledge and decide what to plant, initiate, create, and invent on one's own land, and (5) the right to co-manage or co-govern natural resources with neighboring communities and local authorities.

The Region's Community Spirit Forestlands

Community spirit forestlands are spaces in which villagers practice and maintain their religious values toward natural spirits via traditional rituals. This land (Fig. 1.1) has been recognized over many generations as being owned by the whole community; its management and protection are closely associated with the traditional practices and beliefs of the community, and with the roles of the elders, prestigious people, and clan heads who voluntarily implement its management.

In addition to its spiritual purpose, these forestlands provide resources to ensure the livelihood of households in the community for cultivation, housing, firewood, herbal medicines, and food. From the perspective of the villagers, these community forestland areas have always belonged to them, having been transferred to them by previous generations. However, local communities still lack rights under the law to manage and use these areas.

Central Challenge

LISO sees the central issue facing indigenous ethnic minority peoples in the Mekong region today as two fold: firstly, that of preserving spiritual beliefs and values that are embedded in their relationship to their traditional lands, and which form the basis for their customary laws for governing their land use practices. This is particularly challenging today because these beliefs, values, and practices are different from those that are being promoted globally by large transnational corporations, international financial organizations, and nation-states. However, as the devastating environmental, social, and political consequences of unbridled capitalist development become clear to everyone, the search for an alternative set of values for relating to both humans and nature is becoming increasingly urgent. We believe that the values preserved

[1] This definition was created through consultation among SPERI and traditional healers, spiritual leaders, and ordinary farmers.

Fig. 1.1 Map of community spirit forestlands preserved by LISO efforts

by indigenous peoples around the world are those that we need to adopt to sustain natural forest landscapes.

We recognize, however, that the often-destructive beliefs, values, and practices of corporations, financial organizations, and nation-states are opposed to traditional forms of governance, meaning that these traditional forms have suffered and changed a lot. Communities are also changing as a result. In order to deal with this new context, traditional governance systems need to evolve, adapt, and stand up against this destruction. This would include traditional governance strengthening its voice and becoming better able to resist the overwhelming efforts to change and "modernize" them.

At the next level, one major obstacle to achieving customary law-based land governance is that government land law does not recognize community rights to land; it recognizes only individual households' and individual community organizations' rights to land.

Strategy

Our efforts are directed toward ensuring that indigenous ethnic minorities in the Mekong region retain or recover their rights to their land and that this land continues to be governed according to their customary laws. In this way, the traditional spiritual beliefs and values associated with the land can be preserved as the fundamental guiding principles governing land use. Community land ownership has not been achieved everywhere—individual household ownership is in some cases all that will be allowed by the state; but in all cases, whether the land is granted for communal or individual household ownership, we have ensured that it is governed according to local customary law. We pursue this strategy through a process of "participatory customary law-based community land allocation."

LISO has been able to facilitate accommodations between state legislation and local customary law. Here, local communities have been able to ensure that all their land is governed according to local customary law by institutionalizing their customary regulations for land management and having these legitimized by local government authorities. Contrary to some widely held views, indigenous peoples' customary law is not rigid. It is flexible and adaptable to changing circumstances, situations, and needs. Some indigenous groups (e.g., Hmong) have regular ceremonies for adjusting their customary law to new circumstances.

LISO appreciated that relationships between the indigenous peoples are intimate in terms of shared philosophy, history, and personnel. More than twenty-five years of engagement with indigenous ethnic minority communities in the Mekong region define LISO's added value[2] as an organization.

Methods

These activities aim to persuade local government authorities of the superior knowledge and land management practices of the local villagers, as well as to gain their

[2] One major achievement has been LISO's success in lobbying and influencing the government to make changes to 16 articles of Vietnam's draft Forestry Law, including Article 86, confirming legalization of community ownership of Spirit Forests, Watershed Forests and Production Forests.

support for legalizing the voluntary, community, and customary law-based governance of natural resources by the local people. The first step in this process involves extensive and in-depth community-based research to understand the intimate connection between a people's beliefs and values and their relationship to the land (see Appendix B). The next step involves engaging local village elders in surveying and categorizing the landscape of their villages according to their traditional wisdom, knowledge, and customs. The third step is to empower local villagers as spokespeople and presenters of their indigenous land management practices at farmer-to-farmer and farmer-to-local authority meetings.

Achievements

The LISO Alliance has over twenty years of experience working with highland indigenous ethnic minorities in the Mekong region. During this time, it has built up a very strong network of "key farmers"[3] to act as effective speakers in farmer-to-farmer and farmer-to-authority workshops. They help explain the possibilities, benefits, and advantages of community-based land ownership and customary law-based natural resource management. Positive impacts achieved through the methodology of using farmers as speakers, trainers, and facilitators include the building of farmer confidence and the strengthening of solidarity both within and between villages, thereby enlivening their determination to preserve their culture of living harmoniously with nature.

Another positive impact has been the change in attitudes of local authority staff. It comes as a bit of a shock to them to hear ethnic minority farmers presenting their wisdom and practices of natural resource management and knowledge of the environment, and to see with their own eyes how effective customary laws are for natural resource protection. These shocks have caused them to change their attitudes, from seeing ethnic minority farmers as "backward" and "superstitious" to seeing them as very knowledgeable and capable. The outcome is that local authorities gain complete confidence in the ability of the local people to manage their natural environment effectively according to their knowledge and customary laws. This in turn greatly facilitates the smooth transfer of land titles from state organizations and private individuals to whole communities.

Since 1995, the LISO Alliance and its predecessor organization, Toward Ethnic Women (TEW), advocated for the allocation of land/ tenurial rights to 62,673 hectares of forestland areas to indigenous ethnic minority households and community organizations in Vietnam and Lao PDR. This includes 44,274 hectares allocated to 8,268 ethnic indigenous minority *households* in Vietnam and Lao PDR, and 18,390 hectares allocated to 77 ethnic community *organizations* in Vietnam and Lao PDR.

[3] See Appendix A for a discussion of the role of key farmers in SPERI's efforts to address structural poverty through sustainable development.

In both Vietnam and Lao PDR, land (including agricultural and forestland) is owned by the population as a whole and administered by the state on the people's behalf. Individuals and organizations are granted access to land through land use rights. Communal tenure, which is customary for indigenous communities, is not formally recognized by law. LISO, however, works within these restrictions to bring about formal recognition of indigenous customary law-based land use practices at the local government level and facilitates the legalization of regulations for the communal customary law-based governance of allocated land. In this way, LISO has been able to ensure that land use rights are defined and allocated in a way that conforms to indigenous values of communal tenure. Once the land has been allocated according to customary land use beliefs and practices, communities draw up their land use regulations, which are then officially recognized by local government land management authorities. This allows communities to maintain their own successful ecological land-based livelihoods without fear of outside commercial interventions or land grabs.

The next step is to use the successful case studies of "participatory customary law-based community land right allocation" to lobby the central government for changes in the national law. In this respect, LISO as a Vietnam-based organization has been successful in pushing for the reformed Vietnam Law on Forestry (2017) to recognize community ownership of spirit forests, traditional watershed forests, and traditional production forests. The strategic importance of this achievement is that the 2017 Law on Forestry now stands as a springboard for a new series of actions by NGOs on behalf of indigenous ethnic minority communities in Vietnam. The full backing of the law greatly strengthens the hand of the NGOs and indigenous ethnic minority farmers to act openly in pursuit of their cultural, religious, and livelihood rights.

External Factors

LISOs' work is aimed at bringing about a fundamental change in government attitudes toward indigenous ethnic minority peoples and land management practices. While entrenched attitudes and policies are major obstacles, certain external factors and circumstances have helped in achieving the goal of community land titling and customary law-based land management. Primary among these is the crisis of confidence on the part of local government authorities as to how to protect forests and natural resources from exploitation and degradation. Much of LISO's success has come from being able to supply government authorities with a workable solution to their problem, and one that is desired by and acceptable to the local people.

Obstacles

Not all customary law-based land allocation projects go smoothly. When working with indigenous ethnic minority communities, unexpected things can happen, and unexpected situations can suddenly arise that necessitate a change of action plan. Unfortunately, some donor organizations are insensitive to the need for flexibility when dealing with these situations and may become obstructive or even refuse to cooperate altogether. When this occurs, it is LISO's policy to put the needs of the farmers first and take the necessary actions, even if it requires bearing costs. Efforts, therefore, need to be made to re-educate donor organizations as to the need for greater flexibility in their funding arrangements to deal with these situations. In this respect, donor organizations, as well as local government authorities, become the target groups for necessary changes in attitude.

Vision

LISO continues to work to preserve and enhance its regional network of key farmers and young indigenous leaders that has been built up since the early 1990s by its predecessor organizations—Toward Ethnic Women (TEW 1994), Centre for Human Ecology Studies of Highlands (CHESH, 1999), and Centre for Indigenous Knowledge and Development (CIRD)—and which extends from Vietnam to Laos and Thailand. Our vision is to expand this network to the whole of Southeast Asia and beyond. Connections already exist with Thailand, Laos, Myanmar, Cambodia, Bhutan, Philippines, Indonesia, Europe, and Brazil.[4]

Appendix A: The Dynamic Process of Facilitating NGO Movement to Support Indigenous Ethnic Minority People in the Mekong Region

This process began in 1994 with the establishment of Toward Ethnic Women (TEW), one of the first NGOs to be established in Vietnam. TEW's main focus was upon "Women's Rights" and it oriented its activities toward improving the situation of indigenous ethnic minority women in Vietnam, a section of the ethnic minority population that was most disadvantaged due to the negative prejudices and stereotypes that were held in Vietnamese society in general, by government officials in particular, and even among ethnologists and sociologists, that ethnic minorities were "backward", "ignorant" and "superstitious." In contrast, TEW valued the inherent

[4] Toward Organic Asia, Thailand; Hmong Association of Luang Prabang, Laos; Royal College of Natural Resource Management, Bhutan; MASIPAC, Philippines; GRAIN, Tropical Rainforest Farming, Indonesia; NatureLife International, Germany; Instituto Politicas Alternativas, Brazil.

strengths of ethnic minority women who live close to nature and gain their identity and well-being from nature via their wisdom of medicinal herbs and textile handicrafts to provide for their families' livelihoods. TEW engaged directly with the women to consolidate their strengths and promote recognition of their capabilities, and thereby change community, policymaker, and academic attitudes toward them.

TEW's solution to the problem of structural poverty facing indigenous people was first to break down the feeling of isolation. TEW did this by building up networks of "key farmers" throughout the whole Mekong region (Vietnam, Laos, and Thailand). key farmers are knowledgeable, prestigious farmers willing to learn, do experiments, share, and debate based on their knowledge and integrated and adaptable technologies. They are pioneers in setting up pilot models of farming, animal husbandry, agro-forestry gardens, and use their farms as practical forums to share their experiences and ideas with villagers from different communities, the media, researchers, and policymakers. They engage not only in awareness-raising but also in lobby and advocacy activities for land rights for the community and villagers.

Following its "*Nine-Step approach to structural poverty reduction and sustainable development*" (see Appendix C), TEW brought indigenous ethnic minoriity key farmers together in study tours, workshops, and conferences where they could learn from each other and discover that their experiences of being isolated and marginalized were experiences that others shared as well. The next Step was to build up the confidence of indigenous ethnic minority key farmers to speak out about their concerns. To achieve this, TEW held national conferences at which the farmers could speak about their grievances directly to high-ranking government officials. After this, the key farmers gained the confidence to lobby the government for recognition of their ownership of their traditional land, culture, and identity.

As a result of this work, there emerged a powerful and articulate body of indigenous ethnic minority key farmers who were able to take up the process of organizational and institutional development on their own. As Key Farmer Coordinators and Key Farmer Board Members, different key farmers became actively involved in planning and decision making at the commune, district, and provincial levels. In collaboration with local progressive authorities, and with expert advice from eminent personalities and the support of sympathetic media, they came to play an active and often leading role in lobbying government policy at both the local and national levels on behalf of indigenous minorities, especially in the area of land and forest rights. Some of the key farmers recognized and promoted by TEW became local authority officers at different levels of the official system and were able to introduce the TEW/SPERI method of bottom-up participation into the formal governing system. Some have become parliamentarians, and some have become community entrepreneurs.

In 1999, in the face of government policy of displacing ethnic minority communities from their traditional lands for industrial development (a policy that had a devastating effect not only upon indigenous culture but also upon nature), TEW established the Centre for Human Ecology Studies of the Highland (CHESH) to promote those communities living harmoniously with and nurturing nature through their customary beliefs and practices in their daily livelihood activities. The following year (2000), TEW established the Centre for Indigenous Research and Development (CIRD) to focus upon the rights of indigenous people to live on their traditional land and practice their livelihood according to their indigenous knowledge. From 2000 to 2005, TEW, CHESH and CIRD formed a united front in support of Women's Rights, Nature's Rights and Indigenous Rights to confront the appropriation of indigenous people's natural resources by commercial and political interests.

The regional and national key farmers networks established by TEW developed and expanded across national borders to become Mekong Community for Ecological Trading (MECO-ECOTRA), a regional network of traditional elders and key farmers and a grassroots foundation for traditional civil society organizational and institutional development across national, ethnic, and political borders.

In 2005, to provide a more concerted focus for lobbying government policy for indigenous rights to land, TEW, CHESH, and CIRD were merged into the Social Policy Ecology Research Institute (SPERI). SPERI combined the results of action research undertaken in different ethnic minority communities with policy analysis to provide an evidential basis for lobbying government for policy changes. Major actions were also taken in advocating for a Civil Society Law in Vietnam and in lobbying against Bauxite mining in the Central Highlands of Vietnam.

In general, the direction of SPERI's work has been to oppose the commercial appropriation of indigenous peoples' land and natural resources to defend the cultural and biological diversity of the Mekong region. At the same time, one team of senior members from TEW and CIRD established CIRUM (Culture Identity and Resource Management) to concentrate on Forest and Land Rights for vulnerable groups and on networking for self-sufficiency development. Then, in 2007, CODE (Consultancy on Development) was established from SPERI's Lobby Department as an independent consultancy to focus on bridging and strengthening Public–Private-Civil Society relations in lobbying policy on mining, hydropower, and extractive industries. In its action against Bauxite mining, CODE united 17 independent intellectuals to provide technical evidence on the social and environmental impacts if Bauxite mining was to go ahead. All three organizations share the same philosophy and Founding Board and cooperate extensively.

In the period from 2005 to 2010, SPERI focused on supporting MECO-ECOTRA, with its 6 thematic approaches (1. customary law-based community governance of natural resources; 2. community ownership of spirit forests for bio-cultural diversity preservation; 3. ecological farming for sustainable land use and livelihood security; 4. herbal wisdom for community healthcare and biodiversity preservation; 5. traditional textile handicraft for daily livelihood; and 6. farmer fields school for traditional indigenous knowledge and leadership training).

MECO-ECOTRA places serious attention on capacity building, young leadership development, and pilot actions at 4 levels: (1) household agro-ecological farming enterprising (after a community has received legal Land Right Titles); (2) communal agro-ecological co-governance based on customary law; (3) intercommunity curriculums for exchanging and training of young leadership; and (4) international curriculum for young leadership empowerment and enrichment. Monitoring and evaluation have shown that wherever people over the last 20 years have been devoted to and engaged in their indigenous ways of life living harmoniously with their surrounding nature, worshiping spirit forest, mountain, and water, they have far higher levels of community well-being and happiness.

In order to protect indigenous communities' land from encroachment by commercial interests, *strong community entrepreneurship, community enterprise and community sovereignty* needs to be developed. In order to aid this development, a strong intellectual/eco-agricultural movement is needed to demonstrate to local and national government authorities the benefit for all, in terms of biodiversity preservation and environmental protection services, of continuing that land in indigenous community ownership and management.

More recently, MECO-ECOTRA and its six thematic networks were facilitated to focus more concertedly upon **Y**oung **I**ndigenous **E**thnic **M**inority **L**eadership **D**evelopment **S**trategy **(YIELDS)** for **AG**ro-**E**cological **E**nterprising **(AGREE).** The YIELDS-AGREE strategy will operate at 5 levels with (1) Household Eco-Farming Activists; (2) Community Entrepreneurs; (3) National Intellectual Civil Society Activists; (4) Continental Independent Intellectual Activists; and (5) Global political–ecological activists, with a focus upon promoting local agro-ecological enterprises for community-based livelihood sovereignty for indigenous communities. It was to undertake this transformation that the latest of the LISO Alliance organizations, the Community Entrepreneur Development Institute (CENDI), was established in 2015.

Appendix B: Thirty-Step Method for Claiming Forestland Rights for Ethnic Groups

1. Conduct research, together with local people, on the reality, causes and consequences of landlessness, and find ways for the local people to retrieve land and forest which has been occupied by outside actors.
2. Facilitate briefings and training of key persons and traditional leaders to improve their capacity to negotiate with local authorities and land occupiers.
3. Provide training for key farmers on laws and sub-laws relating to forest land rights, pointing out errors and shortcomings in the current bureaucratic process.
4. Facilitate community-based planning for negotiations, focusing on the role of customary laws, and informing land and forest occupiers of the environmental, social, cultural, moral, and religious outcomes and consequences of the process of land and forest grabbing.
5. Seek consent from land occupiers and local authorities via processes of direct negotiation and criticism.
6. Organize study tours, sharing experiences of methods of community-based land and forest allocation and customary law-based conflict resolution, as illustrated by successful pilot models in Vietnam and Lao PDR since 1995.

7. Facilitate the selection of key farmers, village elders and youths to represent the community, cooperating with professional land and forest technicians and local authorities to participate in training courses and discussions on overlapping land boundaries and conflict.
8. Establish an advisory board for land and forest allocation, which involves traditional leaders, key farmers, representatives of local authorities, and professional technicians.
9. Support this advisory board to work closely with the community to set up regulations, scheduling plans and solutions to the land and forest border overlapping, occupation, and conflicts.
10. Establish a district-level leading board for land and forest allocation, which involves traditional leaders and key farmers, to create opportunities for mutual learning and understanding between formal and traditional systems.
11. Support a taskforce for land and forest allocation in the field, which includes members of the advisory board and the district leading board. This taskforce helps to set up regulations, which correspond to both statutory and customary requirements, aiming at retrieving community traditional land and forests in a peaceful and amenable manner.
12. Organize meeting between local people, experienced key farmers from successful pilot models of land conflict resolution, and occupiers of land and forest.
13. Facilitate a taskforce for land and forest allocation to work with local people (who are selected as household representatives) to survey the land, identifying errors in the maps and borders between households, communities, and enterprises caused by overlapping occupation.
14. Organize training workshops for local people to share evidence of overlapping occupation of community land, traditional sacred forests, watershed forests, herbal forests, and clan forests, and to share experiences from successful pilot models, and find solutions.
15. Organize practical training to assess the capacity of different types of forests on the basis of local knowledge, people's participation, and suitable technology. This step is done before the official measurement and other land allocation procedures.
16. Inform the district chairperson of the reality of land management, overlapping and land occupation.
17. Set up an action plan for the taskforce in the field for lobbying local authorities to make decisions on the allocation of the land affected by overlapping and occupation.
18. Organize training workshops for local people to discuss their rights and obligations in using land and forest allocated by the local authorities.
19. Set up detailed action plans, procedures, land use plan, for forest management in the field, together with local people, representatives of local authorities and technicians.
20. Set up a new set of maps describing borders, land use, and forest management after completing legal procedures and technical, official works.

21. Organize a training workshop for the entire community to help them understand legal decisions, community rights and obligations in the implementation of a land use plan, and the governance of different types of forests.
22. Facilitate discussion among community members on community regulations for the implementation and supervision of land use plans and forest management. This regulation is made on the basis of consent among the entire community and the surrounding communities.
23. Submit a draft of community regulations to the communal authority for monitoring before sending to the district authorities for approval.
24. Document the whole process of land and forest allocation and distribute this among households, communities, and relevant functional offices at district and communal levels.
25. Process data on land zoning, land use planning, forest management, land rights, and forest co-management rights, to register and establish archives at the relevant functional offices at district and communal levels.
26. Organize a ceremony for granting land certificates and rights to the co-management of forest to households, individuals, and community.
27. Support and advise the community management board to set up sign boards with diagrams and regulations on land use and forest co-management, and to clarify border landmarks.
28. Organize workshops at regional and national levels to share the methodological steps for land and forest allocation. Participants in the workshops should consist of local people, local authorities, the media, policymakers, community development organizations, functional technical agencies, and relevant neighboring enterprises and companies.
29. Review and compile records of discussions and analysis from the workshops to arrive at recommendations to send to members of the National Assembly and policymakers dealing with drafting land law. Recommendations are simultaneously updated and posted on SPERI website.
30. Document research and policy analyses of land use and co-management of forests, and socio-economic, environmental, cultural, religious, and political impact indicators, in which land use rights and co-management of forests are the underpinning strategy aimed at strengthening people's confidence, self-determination, and securing their livelihoods on their own land and forests. Integrate documentation of policy analysis of land use plan and co-management of forests into rural development policies, so as to continue lobbying for the land and forest rights of the indigenous ethnic communities in Mekong region.

Appendix C: Nine-Step Approach to Community Empowerment and Institutional and Organizational Development

1. Learn the culture of the local people.
2. Bring the traditional leaders and local authorities together to encourage dialogue.
3. Bring the elders of different ethnic groups together in conferences and workshops.
4. Set up interest groups for bottom-up organizational and institutional development.
5. Build up Community-Based Organizations and Community-Based Institutions to become more professional.
6. Build a system of self-monitoring between interest groups.
7. Offer a platform for key farmers to present to policymakers their own rights-based solutions to problems.
8. Consolidate Community-Based Organizations and Community-Based Institutions for organizational and institutional development.
9. Search for new partners to expand knowledge exchanges and network connections.

Reference

Livelihood Sovereignty Alliance. 2017. *From Community Forest Land Rights to Livelihood Sovereignty and Wellbeing*. Hanoi: Knowledge Publishing House.

Chapter 2
Building Recognition for the Resource Rights of Indigenous Peoples and Local Communities

Antoinette G. Royo, Andhika Vega Praputra, Joan Jamisolamin, and Neni Rochaeni

Highlights

- Recognition of indigenous peoples' and local communities' land tenure rights is a necessary element for securing access to the natural resources they are dependent upon. Recognition of these rights requires affirmative action through multiple approaches by various actors. This includes, but is not limited to, learning and skills sharing, strengthening traditional governance and institutions, strengthening local tenurial capacities, empowering community-based natural resources management, and supporting local conflict resolution mechanisms.
- Government roles include the formal recognition of community rights and allowing access to natural resources. Civic engagement also strengthens rights recognition by influencing decision making, and ultimately by helping to ensure that communities are empowered to manage their resources.
- The success of civic engagement is determined by multiple factors, including good leadership on the part of local administrators, and effective partnerships.
- Civic engagement can produce unexpected negative consequences and these call for effective safeguards (and adaptability) on the ground.

A. G. Royo (✉) · A. V. Praputra · J. Jamisolamin · N. Rochaeni
The Samdhana Institute, Bogor, Indonesia

A. V. Praputra
e-mail: dhika@samdhana.org

J. Jamisolamin
e-mail: joan@samdhana.org

N. Rochaeni
e-mail: neni@samdhana.org

M. Indrawan et al. (eds.), *Civic Engagement in Asia*,
https://doi.org/10.1007/978-981-16-9384-7_2

Introduction

Over the past several decades, inequality and poverty have challenged natural-resource-dependent communities, also known as Indigenous Peoples and Local Communities (IPLC), including those in Southeast Asia (Hall et al. 2011). The region still needs to address development challenges in the social, economic, and environmental dimensions. In most developing countries, especially those with rich natural resources, the state claims ownership of land and forest. Most of the permits or concessions issued by authorities are given to large corporations, limiting—if not eliminating—the access of IPLC to natural resources and putting them in a vulnerable condition. IPLC face procedural challenges if they choose to pursue permits and/or recognition from the government.

Where IPLC rights are recognized and embodied in national laws, such as the Philippines' 1997 Indigenous Peoples Rights Act, the processes and mechanisms should allow fair and just access to natural resources and the benefits they provide. In the bigger picture, these bundles of rights are a good part of human dignity. There needs to be continuous support and accompaniment from various entities, including civil society, to establish the necessary social and environmental movement toward sustainable development in IPLCs.

Issues Prioritized by the Samdhana Institute

The Samdhana Institute began as an association of individual activists, mostly from the Philippines and Indonesia, who dedicated themselves to strengthening the next generation of leaders supporting land rights of indigenous peoples and local communities in the management of natural resources. Problems and issues prioritized for civic engagement by the Samdhana Institute include land rights recognition of indigenous peoples and local communities. This is consistent with the Samdhana Vision: "*A region where natural, cultural, and spiritual diversity is valued and environmental conflicts are resolved peacefully, with justice and equity for all parties.*"

Land is life and lack of land rights is a violation of basic human rights (Gilbert 2013, 115–136), which has resulted in local and indigenous peoples' vulnerability, displacement, criminalization, poverty, and exclusion from the justice system. Lack of land rights has also eroded cultural and spiritual links between nature and humankind, resulting in excessive exploitation of forests and environments. Rights-based approaches in addressing environmental degradation are part of the core competence of the founders and activists of the Samdhana Institute.

Strategies and Approaches

Formed in 2003, the Samdhana Institute now operates in the Philippines, Indonesia, Laos, and other parts of the Mekong region, providing grants, facilitation, and technical support for *local* guardians of natural resources, mainly local community partners and their CSO (Civil Society Organizations) supporters. Thus far, 269 groups in the Philippines; 829 in Indonesia; and 141 in the Mekong region have received Samdhana's support.

Samdhana Institute considers the following strategies:

(1) Long-term and consistent support to IPLC for the assertion of rights and human dignity, natural resources management, resources rights, and conflict resolution.
(2) Urgent response to challenges and opportunities faced by IPLC.
(3) Knowledge and skills sharing of Samdhana Fellows who work directly with indigenous peoples and local community partners in the areas of rights, community empowerment, education, forestry, agriculture, conservation, successor leadership, and others.

Approaches for civic engagement have focused on the inter-related objectives of supporting local communities' self-managed organizations, among them indigenous people's groups, alliances, and organizations.

One approach we have adopted is developing a set of tools to push for recognition of indigenous peoples' land claims. These include the participatory mapping of indigenous territories. The rights-holders negotiate and advocate with third parties using their maps with Samdhana Institute's support and assistance for developing local government regulations to recognize IPLCs' roles and rights at all levels. These efforts, in many cases, also include setting up local organizations and cooperatives for economic empowerment under the framework of multi-sectoral partnerships for community-based natural resource management.

These approaches include:

(1) Facilitating individual and institutional *learning and skills sharing* among actors.
(2) Offering opportunities and support for local community-based actors, including in their efforts to *strengthen traditional governance and institutions.*
(3) *Enhancing and enriching community-based natural resource management (CBNRM)* through the strengthening of local tenure.
(4) Supporting *local conflict resolution* efforts regarding access to land and resources and facilitating peacebuilding for indigenous communities using traditional conflict resolution mechanisms (Fig. 2.1).

The approaches have brought indigenous peoples and local communities together to face license holders with claims over indigenous peoples' land/territory without local consultation, and influential local officials and lawmakers (Fig. 2.2).

Fig. 2.1 Tagbanwa women in Palawan, Philippines, participating in the mapping (Photo by Samdhana)

Fig. 2.2 Head of Merauke regency in Papua, Indonesia, signed the map of Yei tribe (Photo by Samdhana)

Civic Engagement Milestones

Learning and sharing of skills have been fostered at both the individual and institutional levels. At the individual level, commendable local civil society[1] activities involving Laotian ethnic groups have allowed rural and indigenous youth (many of whom are unaccustomed to and reticent about public speaking) to speak up. The youth are encouraged and supported to volunteer for a year with selected partner organizations. Through engaging in community work, youth gain skills and experience they can in turn use for personal and community growth and development.

Learning and sharing of skills happen through direct engagement in a range of community activities. Local groups gain new skills and share skills among their members through initiatives[2] such as women-led conservation and income-generating activities along the Mekong River. For example, a bird-watching program was developed in Lao PDR primarily to protect the riverbank and island habitats of endemic and migratory birds. At the same time, the project area was opened as a tourism destination to augment the income of the women living in this area. The local women formed a committee and worked to establish policies to manage local ecotourism activities and protect the bird habitats and natural ecosystems within which they each function.

Supporting traditional governance and institutions in indigenous communities strengthens local leadership. In Mindanao and many other parts of the Philippines, grant projects served only as its enablers. In the process of project implementation, local leaders themselves manage their activities using culture-based policies and practices. Projects provide increased spaces where community members come together to discuss their visions and plans for their traditional territories. We encourage community partners to mobilize their own human and natural capital as their counterpart to grants. In Northern Mindanao, Philippines, the Higaonon people undertook their reforestation initiative using endemic species countering government promoted non-native species, as a counterpart to their livelihood project. An exemplary local initiative comes from the Manobo community in the Agusan Marsh, which devised the local policy for "no-fishing zones" and "no electric fishing."

An example of strengthened tenure comes from a civil society partnership[3] in Central Sulawesi, Indonesia, that eventually granted the Marena community its first customary forest, about 1,161 hectares of the 1,806.5 hectares proposed. Since the early 2000s, the indigenous community, with support from civil society groups, has attempted to restore local tenurial rights from a provincial government-owned

---

[1] Huamjai Asasamak Association, which established Laos' first national youth volunteering program in 2007, engaging volunteers from rural areas and ethnic minority communities, and particularly women, aged between 18 and 25.

[2] Agro-Forestry Development Consultant in cooperation with the Lao Women's Union.

[3] This partnership was fronted by local NGOs, such as the Central Sulawesi branch of the Indigenous Peoples' Alliance of the Archipelago (AMAN), Bantaya, Yayasan Merah Putih, and Yayasan Pendidikan Rakyat. It was supported by the Agrarian Reform Working Group of Sigi and, at the national level, by the Office of the Presidential Staff (KSP).

concession company in the village. Support for village land use classification came from Samdhana in 2014. The transformation started with a participatory mapping of the high carbon value conservation landscape. The initiative to redress agrarian reform and social forestry soon received support from the new regent (head of regency) at the end of 2016, who also helped facilitate the write up of a road map for agrarian reform. The movement eventually received support from the offices of the Indonesian Ministry of Environment and Forestry and the Presidential Chief of Staff. A decree from the Ministry of Environment and Forestry certifying the assignment of customary forests for the Marena community was granted in 2017.[4]

The measure to enhance and enrich CBNRM includes various innovative activities, such as in Northern Mindanao,[5] where Samdhana supported two start-up initiatives. The first phase focused on establishing diversified and integrated farming systems. The second phase developed value-added products, including the processing of lemongrass oil and the growing of cash crops such as strawberries. Hands-on-learning complemented the process. The livelihood programs of Pamalihi and Daraghuyan also incorporated important landscape management activities whereby detailed land use activities were identified and delineated and zonation was applied (of zones for "production," "buffer," and "forest"). Planting of crops and planned reforestation activities ensure productive efficiency of land and ensure longer-term livelihoods for the Higaonon and Bukidnon tribes. Planning was also intended to create symbiotic relationships; for instance, first they plant indigenous tree species to provide shade for the planting of tissue-cultured abaca (a banana species native to the Philippines). Experiences with Pamalihi and Daraghuyan communities show how indigenous communities can build their capacity and confidence through embarking on livelihood endeavors.

Resource conflicts occur frequently in some localities, but even so, they may be approached with **local conflict resolution mechanisms**. In the case of the Menuvu tribe in Mindanao, the community has been fraught with restricted economic opportunities, coupled with competing land claims from the Maguindanawon/Moro communities that surround the Menuvu's territory. For more than a decade, the Menuvu tribe struggled to gain formal government recognition over their territory. Notwithstanding the lack of support from government agencies, an organization representing the Menuvu people undertook surveys to establish the boundaries of their traditional territories in 2007. In 2011, armed encounters erupted between the Menuvu and Moro groups over the boundary conflict. Samdhana supported a fact-finding mission to help re-establish peace between the indigenous communities involved. The mission helped to make concrete recommendations to the local government and the National Commission on Indigenous Peoples for immediate action and long-term solutions to this recurring conflict. Samdhana also supported the facilitation of local resolution

---

[4] Decree of the Minister of Environment and Forestry SK.1156/Menlhk-PSKL/PKTHA /PSL.0/3/2017.

[5] The southern Philippine island of Mindanao is a place where precarious situations have erupted over the issues of land tenure, livelihood, and security.

and peace mediation between the conflicting tribes. By providing an equal opportunity for both communities to participate in the oil palm enterprise, agreements were reached. Both communities identified common areas to start the oil palm plantation, and these areas will serve as the boundary between their respective lands. The leaders from both communities agreed to settle their territorial conflict and cooperate to allow the oil palm enterprise to start. This effort engaged the congressional district representative, whereby the local government unit supported the launching of the initiative.

In conflict and poverty affected areas, Samdhana's work with the Banwaons in Agusan del Sur (Mindanao) may be considered a good milestone. The tribe has long been enmeshed in a protracted war waged by revolutionary groups against the government. The Banwaons have also long resisted the logging of their sacred forests, which until today are abundant in dipterocarp hardwoods and rattan. Because of volatile security, the core of their rainforest remained intact, but a wide area is invaded by falcatta (*Paraserianthes falcataria*) as a result of massive logging plantations in nearby areas. Through facilitation support (meetings, agreements, and linkages with the private sector and government officials, as well as assistance in the development of planning activities) from Samdhana's Indigenous Peoples Support Fund, the tribe agreed with private investors to harvest falcatta and develop the logged-over areas for food production. The profit is set aside to fund the implementation of the Banwaon's unified Ancestral Domain for Sustainable Development and Protection Plan (ADSDPP) and provide for livelihood, education, health care, and forest regeneration and protection. The Samdhana community development facilitators helped the community establish their structure and system to manage their books and business partnerships. Part of the process was to review the community plans to prioritize the needs and concerns of the different *banwa* (clans). It aimed to set up a trust fund and a social community fund as mechanisms for managing their income. An intense process of balancing the interests of the more than 100 *banwas* while creating a genuine commitment to peace and cooperation is required for this to succeed and benefit everyone. The conduct of a major traditional conflict resolution (*husay*) and peace pact among their tribe members who were internally at war[6] was key to turning the situation around and making it conducive for a tribal enterprise. The *husay* was done after decades of estrangement between warring clans. It was a very significant moment when they decided to bury their age-old conflicts and restore good relationships, through the sacrifices and prayers offered in the ritual.

[6] In their history during the revolutionary war (between the 1970s and 1980s), members of the same tribe took different sides or were recruited by either the military or the revolutionary armed groups. In the end, members of the same tribe were killing one another, resulting in conflicts between families and clans.

Discussion

The success of outcomes induced by civic engagement was influenced by several internal and external factors. Internal factors include good shared vision and thought leadership, as well as competent management systems. An example of good leadership in Indonesia comes from Papua's Jayapura regency, where the regent initiated the passing of a regulation in 2016 recognizing the existence as well as rights and responsibilities of the indigenous peoples.[7] Similarly, in Biak Regency (West Papua province), the forest management unit could develop its protection functions mainly because the regent was able to change the mindset of the government bureaucracy toward finding creative solutions.

A key factor limiting desired changes is the extent of participation of all members and activists in the work to realize the vision. Tambrauw Regency Government (in West Papua province),[8] with a well-balanced ridge to reef feature, scored a milestone when it proclaimed itself a conservation regency.[9] However, a national government program supporting housing moved many local community members to build as many houses as they could resulting in an expanded forest clearing (for land and timber).

External factors that help achieve desired changes include strong alliance-building among civil society organizations (who share the same principles and values), good timing of action, as well as multi-stakeholder (government, business, CSOs) participation and buy-in. The establishment of customary forests in Marena owed its success to a large extent on civil society collaborations (internally and with the local government). Factors that limit desired changes include perverse incentives and bad governance, changing regulations, and turnover of good government counterparts. One example is the recent policy allowing the local market in Merauke to receive a steady supply of instant/packaged food. The local fried sago and sago meatballs were pushed aside. As a response, Samdhana and local partners supported local champions and role models to encourage a return to local food, with approaches including training the local indigenous people in business management techniques.

The Samdhana partnerships put into context attitudinal changes, social practices, and policies among the participating stakeholders (see Table 2.1). Still, the changes take many years to accomplish and are often beyond the normal cycles of time and funding required by development partners.

Changes may be measured by several approaches including combined, results-based, and asset-based approaches (involving human, social, physical, natural capital); development of a joint set of indicators to use for joint monitoring; using innovative tools like Appreciative Inquiry, which is a form of action-research to promote social innovation as defined by Cooperrider and Srivastva (1987).

---

[7] Jayapura Regency Regulation No. 8/2016 regarding Customary Villages.

[8] Of the 1.16 million hectares of land territory, 80% have been decreed as protection forest and conservation areas.

[9] This occurred in 2018, through Tambrauw Regency Regulation No. 5 / 2018 regarding the establishment of Tambrauw as a conservation regency.

Table 2.1 Desirable behavior changes, social practices, and policy-level indicators

Objectives	Key approaches	Indicators
Recognize Rights (bundle of rights—possession, control, exclusion, enjoyment, disposition)	Free-Prior-Informed Consent of indigenous peoples and local communities who have long-term and cultural-historical land rights/claims	Passing of regulations that recognize diversity of ethnicity, culture, or basic rights of indigenous peoples and local communities
Access to opportunities	Making resources/services accessible to indigenous peoples and local communities	Regulations not only recognize rights but also provide resources/support for these rights to be realized
Control over territories	Technical capacity for mapping and determination of boundaries	Support system for registration of rights and tenure agreements granted
Wellbeing enhanced	Festivals and celebrations of life and bounty	Institutions supported, products, financing, and marketing support systems developed

Unexpected negative consequences for civic engagement have occurred, such as dependency, erosion of trust when promises are not delivered on, short-lived achievements, and occasionally delivery incompetence. We have seen in the field instances of CSOs competing (rather than collaborating). In some communities, the benefits were retained by members of the ruling clans. In others, the group coordinators and not the members were doing the bulk of the work. In some localities, the communities began to expect an honorarium for each meeting, or went after the freely distributed farm spades, but did not tend the nurseries or plantations. In some cases, poor and disadvantaged status may not be the most important selection criteria of the target beneficiaries. Instead, projects should consider the needs and passions of the target beneficiaries. This lesson is illustrated by the case of the ginger planting project in Sulawesi that was later abandoned due to a lack of interest on the part of the communities.

We can address such negative consequences by introducing systems for good resource mobilization, accountability, adaptive management, and legal and policy development strategies before initiating projects.

Collaborative learning and skills sharing are necessary to secure and strengthen local communities' land tenure rights as they empower and strengthen traditional governance institutions. The strengthened capacity of communities will, eventually, create a mechanism to support conflict resolution regarding natural resources. Additionally, communities with improved capacity to manage their natural resources will benefit from strengthened livelihood systems and, hopefully, form new solutions to the social, economic, and environmental challenges they face.

References

Cooperrider, David L., and Suresh Srivastva. 1987. "Appreciative Inquiry in Organizational Life." *Research in Organizational Change and Development* 1: 129–169.

Gilbert, Jeremie. 2013. "Land Rights as Human Rights: The Case for a Specific Right to Land." *International Journal on Human Rights* 10, no. 18: 115–136.

Hall, Derek, Philip Hirsch, and Tania Li. 2011. *Powers of Exclusion: Land Dilemmas in Southeast Asia.* Singapore: NUS Press.

Chapter 3 The Heartware of Ecological Sustainability in the Asian Context

Dicky Sofjan

Introduction

Sustainability means many different things to different people. Its definition has been extended to mean all things related to climate change, global warming, green technology, smart city planning, and many others. As such, it is an elusive concept and has many facets to it. Scholars have defined sustainability as balancing "biological requirements" of the planetary system and the "anthropocentric needs" of human beings (McDonach and Yaneske 2002). The main argument, as articulated in Lynn White's work (1967), is that "what people do about their ecology depends on what they think about themselves concerning things around them." The problem, as Nasr (1989) maintains, is that modern humans have developed an inability to appreciate nature and aesthetics due largely to the "externalization of knowledge" and the loss of "sapiential wisdom," which have caused humans to ultimately forget their place on this planet.

The body of literature on sustainability covers three predominant approaches: hardware, software, and heartware (Harashina 1996; Nakamura and Rast 2011; Mohamad et al. 2015, 2018). Hardware involves the scientific study of ecology, and how to deal with it by employing technology and instrumentation. This approach involves hard data and tools to mitigate the problems associated with the changing nature of ecology and the negative impacts of the anthropocene, which is regularly understood as the shifting geological constellations caused by human activity and industrial exploitation. Meanwhile, software refers to the laws, regulations, and policy frameworks imposed by authorities in ensuring that environmental safety and ecological sustainability are maintained and practiced. In short, both approaches serve to uphold the principles of sustainability.

D. Sofjan (✉)
Indonesian Consortium for Religious Studies (ICRS), Gadjah Mada University (UGM), Yogyakarta, Indonesia
e-mail: dickysofjan@ugm.ac.id

M. Indrawan et al. (eds.), *Civic Engagement in Asia*,
https://doi.org/10.1007/978-981-16-9384-7_3

Unlike the hardware and software approaches, the heartware relates to the individuated, non-tangible, and inner dimension of the drivers of sustainability. It points to values, faith, religious convictions, and spiritual practices in all the things beyond the materiality of ecology and the workings of ecological governance. It is based on a sense of spiritual interconnectedness and a heightened awareness of the role of human beings as "stewards of the earth."[1]

If heartware is neglected, then ecological sustainability transition strategies would be reduced to scientific endeavor and technocratic affairs, with no consideration for the most fundamental and central aspect of human existence. Human beings are driven not only by survival instincts or material benefits, nor do they always respond positively or as expected to science, regulatory frameworks, and public policies. Humans have a yearning to search for meaning in life and connections that go beyond their existence and bodily experience. Among them is a sort of planetary consciousness, a sense of interconnectivity, and a desire to do good for the betterment of humanity and generations to come based on the very core of their spirituality. Hence, the idea of "transcendence," defined as "a way of human knowing" and "going beyond the here and now," becomes relevant (Duara 2015).

Such heightened awareness or consciousness is derived principally from the manifestation of one's values, faith, religion, and spiritual mindfulness. It is the "psychic foundations" that link all beings with the planet and beyond which could save us all from "man's exploitative attitude toward nature" (White 1967).

> What people do about their ecology depends on what they think about themselves in relation to things around them. Human ecology is deeply conditioned by beliefs about our nature and destiny, that is, by religion … The spirits in natural objects, which formerly had protected nature from man, evaporated. Man's effective monopoly on spirit in this world was confirmed, and the old inhibitions to the exploitation of nature crumbled. (White 1967, 1205)

In the absence of faith, religion, and spirituality, no genuine awareness and commitment to sustain life and conserve the ecological surrounding would or could ever emerge, as anthropocentrism would take precedence over the long-term biological needs of the existing ecological system. Tucker and Grim maintain that:

> Religion connects humans with a divine or numinous presence, with the human community, and with the broader earth community. It links humans to the larger matrix of mystery in which life arises, unfolds and flourishes. (2001, 14)

Yet, insofar as the heartware is concerned, no other stance is as powerful as the one made recently by the Pope as the world's Catholic leader. The "Encyclical Letter *Laudato Si'* of the Holy Father Francis on Care for Our Common Home" (Francis 2015) directs the world's attention to the fact that the Catholic leadership is making bold steps toward thinking beyond the here and now and toward working out a strategy for how to sustain life on this planet. The Encyclical Letter suggests that we should do away with "pollution, waste and the throwaway culture" and insist that the "climate [is] a common good." It further articulates the "decline in the quality of human life

[1] In Islam, the Quranic lexicon used to refer to this status of humankind on earth is encapsulated in the phrase "*khalifah fii al- 'ardhi*" (see the Quran, 2:30).

and the breakdown of society" and the clear existence of "global inequality." In short, the Encyclical Letter maps out the dire state of our planet and the current problem of humanity, insisting that humanity could and should utilize religious and spiritual resources at our disposal.

After defining the heartware of sustainability, the following section will elaborate on how the heartware is applied by local communities in five Asian countries, and what one could learn from each of the respective community experiences.

Learning from Asian Communities

Asia is home to a vast number of faiths, religions, and spiritual practices. Despite the onslaught of modernization and global capitalism, some Asian communities remain faithful to their core values, beliefs, religions, and spirituality. This has been manifested in the many prevailing rituals and community festivities that ultimately define the heartware of ecological sustainability in Asia.

This article is based on reflections and lessons derived from the Asian Public Intellectuals' (API) regional project, which took place between 2008 and 2014. The regional project—entitled "Community-based Initiatives toward Ecological Balance"—involved over 250 public intellectuals, academics, researchers, civil society activists, and artists in Southeast Asia and Japan. The research locus was set at five sites, i.e., Kali Code in Yogyakarta (Indonesia), Tasik Chini in Pahang (Malaysia), Khiriwong in Nakhon Si Thammarat (Thailand), Batanes in Batan (Philippines), and Biwako in the Kinky region (Japan).

In each respective site, API fellows, both local and foreign, collaborated with their local community counterparts throughout the duration of the regional project. Taking various shapes and forms, many of the collaborations with the local communities blossomed into genuine friendships that have continued to this day. In the regional project, local API fellows were tasked with collaborating with the local communities to host a week-long site visit for visiting foreign fellows. In most instances, many of these site visits included local government officials and involved stakeholders such as university academics, researchers, NGO activists, musicians, artists, small business owners, and other relevant partners.

At the end of the regional project, the API fellows produced a beautifully designed and edited volume entitled *Living Landscapes, Connected Communities: Culture, Environment and Change across Asia* (Vaz and Aphinives 2014). The volume comprises 372 pages of around 50 evocative articles, academic analyses, and poems sprinkled with striking pictures depicting the five localities. Following the edited volume at the regional level, API fellows in the five participating countries also published books about their respective sites in their national languages. This scheme was deliberately designed not to promote the work of the API in these localities, but rather to highlight the work of the local communities in ecological sustainability and to raise awareness of certain aspects of their pertinent work on the ground.

Table 3.1 Ecological features and heartware elements in five Asian localities

Country	Site	Ecological Features	Predominant Religion	Heartware Elements
Indonesia	Kali Code	Downstream/upstream river, dense urban area	Islam	Mosques and Merti Code ritual and festivity
Malaysia	Tasik Chini	Lake, forestry, incursion of palm oil plantation and extractive industry	Islam	Myth of the *Naga* (Dragon Spirit)
Thailand	Khiriwong	Mountain village, upstream river, *suan son rom* (orchard)	Buddhism	Commemoration of the Great Flood, Boat Carrying Festival, role of Buddhist monks
Philippines	Batanes	Coastal area, natural port, fishing village and rolling hills	Catholicism	*Mayvanuvanua* ritual and shaman
Japan	Biwako	Forestry, upstream river, agriculture, downstream lake	Shinto and Buddhism	Forest and river gods, *Mukugawa* Spring Festival

In addition to books and printed materials, the regional project also produced six research documentaries by engaging a roving film crew from the Manila-based Center for New Cinema (CNC).[2] The documentary making was led by an award-winning film director from the Philippines, Nick Deocampo, who happens to be an API fellow. The film crew recorded hundreds of hours of footage and recorded interviews with informants and stakeholders in all five sites. The regional documentary, entitled *Cross Current*, was intended to be one of the main marketing tools for the API Community in the years to come. Apart from *Cross Current*, CNC also produced five other documentaries at the site level, highlighting the achievements and challenges faced by each respective local community.

In all five sites, local communities in alliance with like-minded people and groups have taken the leadership role in environmental sustainability efforts. Through the regional project, it was found that the heartware of sustainability has been played out vis-à-vis the scientific endeavors conducted by university researchers and the policymaking framework developed by the local authorities. Admittedly, in some places, the heartware is more salient than in others. The following section examines the role of the heartware in each of the localities and shows how it manifests itself in the work of the respective local community (Table 3.1).

Above is a summary of the ecological features and heartware elements in five Asian localities. Below I will explain each of them in turn.

[2] View the documentary trailers on CNC's website: http://centerfornewcinema.net/api-docs.

Kali Code

Kali Code is one of three rivers that run through the city of Yogyakarta, Indonesia. The community in Kali Code could perhaps provide a prime example of how heartware is optimally utilized by the local community. Many of the environmental activists in the riverine community are religious and active leaders in their faith communities. Some even operate from their religious houses of worship, such as mosques, and have significant influence over their faith community. More importantly, the Kali Code leaders have been skillful in framing their commitment to ecological sustainability within the context of their religious convictions.

The group which the API fellows engaged, *Pemerti Kali Code* (Preservers of the Code River), has a reputation for being well connected to an array of stakeholders, including local governments, non-government/civil society organizations, higher education institutions, research think tanks, the media, and others. Due to their intense and deep interaction with many stakeholders, *Pemerti* has been quite successful in resource mobilization and garnering support for their causes, including from government, national, and international universities as well as funders such as The Nippon Foundation. The media, too, has been a good friend of Kali Code for many years, reporting on the community's activities in ecological sustainability. In fact, just recently, the leader of *Pemerti*, Totok Pratopo, received the *Kalpataru*, a prestigious national medal for environmental care, protection, and conservation from the Indonesian government.

Through API fellows' engagement with the Kali Code community, it was apparent that not only were hardware and software important to their work, but also that the heartware has been equally instrumental in the way *Pemerti* and other groups have mobilized the local riverine community to be aware of the need for a well-managed river, particularly in the case of one that is prone to disasters such as annual flooding.

In Kali Code, an event takes place each year to commemorate and celebrate the blessings of water deriving from the seven spring waters that are the source of the river. The ritual, called *Merti Code* (Preserving Code), is participated in by all three sections of the riverine community located in the north, central, and southern areas as well as local government leaders and a wide range of other relevant stakeholders. The community ritual is central to the preservation of Kali Code, as it builds awareness of the centrality of water as a vital element in life. The water ritual, followed by month-long festivities and activities, lends credence to how the heartware could support ecological sustainability.

Tasik Chini

In the context of Tasik Chini, Malaysia, the heartware revolves around the myth of the *Naga* (Dragon) that resides in the "seas,"[3] and all the related beliefs, customs, and

[3] In the distant past, the *Orang Asli* tended to call the big lake "the seas."

traditions that the *Orang Asli* (Original People) have preserved until today. For them, preserving their forest and lake has direct consequences for their survival as a community and people. The forest surrounding the Tasik Chini not only provides them with food and shelter, but also with an array of herbal medicines and aphrodisiacs such as the famed *tongkat Ali* (*Eurycoma longifolia*) and *kacip Fatimah* (*Labisia pumila*).[4]

Unfortunately, Tasik Chini, located in the state of Pahang, had already been negatively affected by overdevelopment long before the existence of API's Regional Project. It started with the building of a weir to essentially control the water level coming from the Pahang River. As a result, water flow in the lake was no longer natural and the water quality deteriorated rapidly, causing the loss of nutrients and increased levels of toxicity, which ultimately depleted the fish and other organisms in the lake. Slowly but surely, Tasik Chini is now awaiting its death, as the proliferation of *ekor kucing* (cat's tail or *Cabomba furcate*) at the bottom of the lake can no longer be controlled, suffocating anything that is under and above them. Even the pink tulip, once the pride of Tasik Chini, was already at around 5% when the Regional Project started to take off.

This is not to say that the local community in Tasik Chini did not resist the ecological destruction. However, much of the heartware left in the community was made to serve the needs of the surrounding palm oil plantation and the extractive industry that required hard labor from the neighboring villages at the expense of its once beautiful lake and lush forest. Many initiatives and ideas by the *Orang Asli* have sprung up, especially in the field of alternative livelihoods and herbal medicines amid the growing eco-tourism industry. Sadly, the eco-tourism does not stand a chance against the past and present industrial policies of the state of Pahang, which could completely destroy the natural beauty and biodiversity that Tasik Chini could offer.

In spite of the negative development and the unbridled capitalistic tendencies of the local government, the *Orang Asli* in Tasik Chini remain loyal to their traditions and customs that have been passed down through generations. The question is whether the collapse of Tasik Chini would leave anything for the *Orang Asli*, or whether the local community would be able to outlive their unsustainable ecological system.

Khiriwong

A set of villages in the San Laka district in the province of Nakhon Si Thammarat, Khiriwong (officially, Baan Khiri Wong) is nestled in the valleys of a mountain range in southern Thailand. It prides itself on its ancient history, lush forest, cool air, fresh produce, and beautiful river that flows directly from the mountains to the downstream villages, making the soil very fertile for growing fruit and vegetables.

[4] Interestingly, both names of Ali and Fatimah are derived from the cousin/son-in-law and daughter of the Prophet Muhammad of Islam. In Islam, Ali and Fatimah are symbols of fertility and vitality. They bore many devout sons and daughters, who are revered until today. In Shia Islam tradition, the Prophet Muhammad, they and their offspring are called *Ahl al-Bayt* (People of the House of the Prophet) and subsequently "*al-Ma'sumiin*" (The Infallibles).

Due to these superb natural endowments, city dwellers flock to the village, coming to seek sanctuary from Thailand's densely populated and smoky cities.

One of the main draws to Khiriwong has been its forest orchards (*suan som rom*) located in the upland forests. People gather to learn about the cultivation of fruit and vegetables, and to bring some of this produce back to the city. Some of the most popular fruit in Khiriwong include purple mangosteens (*Garcia mangostana*) and durians (*Durio zibethinus*). The mangosteen is particularly central, as villagers turn the mangosteen rind into natural soap, which is also sold to tourists. Meanwhile, the annual durian festival is one of the most well-attended community events in the area, drawing durian lovers from all corners of the country and region.

Given its abundant and sustainable natural resources, it is no wonder that Khiriwong villagers are so grateful for their "self-sufficiency," which seems to be an effective mantra for the local community. Rituals and community festivities are a common sight in Khiriwong, with plenty of food to share among members of the community and also visitors who flock to and participate in these events. In all these rituals and festivities, the Buddhist temple holds a special place and significance not only as a religious space but also as a community center where people from different walks of life converge and interact in the spirit of togetherness.

One particular festival, the boat carrying festival, brings the community back in time by recalling the barter and kinship systems developed between the upstream and downstream villages. It is preceded by a set of prayers and chanting led by Buddhist monks, which are then followed by a community feast for everyone involved. But, like Kali Code, Khiriwong is also a disaster-prone area, which stems from the overflowing of the river during the heavy rain season. Residents, therefore, commemorate the Great Flood every year. In this context, rituals and festivities in Khiriwong have been instrumental in ensuring community resilience and ecological sustainability.

Batanes

Located in the northern tip of the Philippines, Batanes is an archipelagic province with breathtaking scenery and picturesque topography, overlooking the South China Sea in the west and the Philippines Sea in the east. The collection of islands in Batanes is simply a magical place that has been inhabited by the legendary indigenous group, the Ivatans, a group of Austronesian descent that goes back at least 1,000 years. Despite its exquisite beauty, heavy downpours, strong winds, and typhoons are a regular climatic feature of Batanes. Fortunately, the Ivatans are a highly adaptive and resilient community with a deep sense and knowledge of their surrounding ecology.

For a long time, their lifestyle and livelihood have been determinedly committed to disaster preparedness and mitigation. With the current influx of tourists, creeping capitalism, and technological development, some aspects of Batanes life have faced substantial challenges, as economic growth and social structure are changing rapidly to accommodate modern practices. Furthermore, the development of the local tourism industry has effectively turned the Ivatans into what Florentino Hornedo disparagingly calls "living artifacts" for outsiders to observe.[5]

Despite these developments, some things never change. Ivatans still hold firmly to their indigenous beliefs, which are tied to their ecological awareness and spirituality. In the village of Diura, a *mataw* fishing village in Mahatao, there exists a *vanua* (natural port) at Valugan Bay, where fishermen come and go to the ocean. The *Mayvanuvanua* (the "opening" of the seas) ritual has over many generations become illustrative of a kind of impeccable fusion between spirituality and ecological sustainability. The *Mayvanuvanua* marks the annual opening of the traditional fishing season to catch *dorado* or *arayu* (*Coryphaena hippurus*). The season shifts from time to time, depending on the period of the typhoons. The ritual itself is led by a local shaman and involves many sacred rituals and taboos for the fishermen and their families as well as visitors.

With the existence of such customs and traditions, transferred from one generation to the next, the Ivatans can develop a close-knit community and maintain a decent livelihood during the season, while engaging in ecological sustainability. The challenge comes from the neighboring countries that engage in poaching and illegal fishing in the waters of Batanes using modern, giant vessels, which could deplete the living resources and ultimately tip the balance.

Biwako

The Biwako area in Japan is known for its *satoyama*, a traditional system of resource management, made famous by a documentary entitled *Satoyama: Japan's Secret Water Garden* (NHK 1999/2004), which was directed by Masumi Mizunuma and narrated by David Attenborough. Being Japan's largest lake, Biwako holds not only significant ecological but also historical significance. It is said that the lake, which provides water to over 15 million people, was formed[6] some four million years ago when dinosaurs still strode the region.

The villages of Harihata and Mukugawa, where the site visits were conducted, are located in the upstream area of the Ado River. Both villages are surrounded by natural and planted cedar forests. It is no wonder that previously Harihata used to be one of the main sources of exported timber. However, with the rise of cheap timber from Southeast Asia, prices started to drop, and the industry fell on its knees. Meanwhile, Mukugawa used to be a charcoal producer, which enabled the local community there to thrive. Due to the decline in the timber and charcoal industries and the resultant economic stagnation, the area was hard hit by rapid urbanization, leaving aged people to sustain life, manage the forest, and tend to the paddy fields.

Notwithstanding these challenges, the residents' concept of "living with nature" remains strong and is nurtured through rituals and community festivities. These rituals and festivities serve as a reminder to the community about the need to preserve

[5] Statement made to API fellows during the site visit in Batanes in 2011.

[6] Nature Documentary hosted by David Attenborough and published by NHK in 1999. https://www.youtube.com/watch?v=94qMbmgWcO4.

nature, consume in moderation, and not get involved in over-exploitation. Examining the *satoyama* from a social-ecological perspective, Suzuki (2014, 70–83) argues that such reminders are derived from locally generated knowledge, beliefs, and reverence for the mountain and river gods.

These residents instinctively know that both the forest and river are vital sources of life, and that to degrade them would mean the extinction of the community. The Mukugawa festival, which coincides with the spring equinox, also begins with honoring these gods as well as their ancestors. It brings the community together to tap into the same spiritual frequency, which bolsters their commitment to ecological sustainability.

Conclusion

Values, faith, religion, and spirituality, as the main elements of the heartware approach, are proven to have direct consequences with how local communities perceive and practice ecological sustainability. Based on research and engagement in five Asian local communities, the heartware of sustainability was present, except in the dire case of Tasik Chini, Malaysia, where a "crisis of transcendence" (Duara 2015) seems to be unfolding due largely to unbridled capitalism and the over-exploitation of natural resources.

For the most part, local communities in these five localities in Southeast Asia and Japan have displayed a strong affinity with their surrounding ecology and enthusiasm in engaging the sacred through rituals and festivities. The inclination toward ecological sustainability is clearly and deeply conditioned by their faith, religious conviction, and spirituality, which altogether have maintained, expanded, and cultivated their heartware. This is manifested by the local communities' commitment, initiatives, and activism toward creating a balance between nature and culture.

This inner dimension perspective on the heartware of ecological sustainability is "still a working concept" (Mohamad et al. 2015) and needs further elaboration and theoretical underpinnings. Having said that, the heartware approach is relatively stable and seems more durable and dependable in contrast to the more rational hardware approach and the more predictable software approach. Rituals and festivities are commonly transferrable from one generation to the next through collective memory and consciousness, which foster respect and appreciation for the ecology. The heartware approach is also dependable because faith, religion, and spirituality form a specific worldview that sits comfortably with ecological sustainability.

I believe that the emphasis on heartware provides a more holistic view of human agency in light of the changing natural landscape, habitat loss, and species extinction caused by modernization, global capitalism, and climate change. Faith, religion, and spirituality can shed light on the more reflective side of ecological sustainability, as they could keep human avarice, greed, and overconsumption at bay (Figs. 3.1, 3.2, 3.3, 3.4, and 3.5).

Fig. 3.1 Tasik Chini—API fellows ride a boat at the break of dawn in Tasik Chini to explore the degrading ecosystem (Photo: Dicky Sofjan)

Fig. 3.2 Kali Code—Mount Merapi erupts in late 2010 located in the northern area of the Yogyakarta province, where the Kali Code community dwells. The cold lava travels along the Kali Code engulfing the houses along the riverbanks (Photo: Dicky Sofjan)

Fig. 3.3 Khiriwong—villagers carry a boat to the river during a ceremony after being blessed with holy water and incantations by local Buddhist monks at a nearby temple (Photo: Dicky Sofjan)

Fig. 3.4 Biwako—API fellows help plant, cultivate, and harvest paddy field in Mukugawa, one of the villages along the Ado River in the upstream region of Lake Biwa (Photo: Dicky Sofjan)

Fig. 3.5 **Batanes**: Rakuh A Payaman, a communal pastureland, oversees the scenic coastland of Batan. An old, leaning tree symbolizes the perseverance of the local and legendary indigenous community of the Ivatans against the harsh realities of the archipelago (Photo: Dicky Sofjan)

References

Duara, Prasenjit. 2015. *The Crisis of Global Modernity: Asian Traditions and a Sustainable Future.* Cambridge, UK: Cambridge University Press.

Francis, Pope. 2015. "Encyclical Letter *Laudato Si'* of the Holy Father Francis on Care for Our Common Home." http://w2.vatican.va/content/francesco/en/encyclicals/documents/papa-francesco_20150524_enciclica-laudato-si.html.

Harashina, Sachahiko. 1996. "Environmental Planning on Urban Level." Discussion Paper 96-6. Department of Social Engineering, Tokyo Institute of Technology, Tokyo.

McDonach, K., and P.P. Yaneske. 2002. "Environmental Management Systems and Sustainable Development." *The Environmentalist* 2: 217–226.

Mohamad, Zeeda Fatimah, Affan Nasaruddin, Siti Norasiah Abd. Kadir, Mohd Noor Musa, Benjamin Ong, and Nobumitsu Sakai. 2015. "Community-based Shared Values as a 'Heart-Ware' Driver for Integrated Watershed Management: Japan-Malaysia Policy Learning Perspective." *Journal of Hydrology* 530: 317–327.

Mohamad, Zeeda Fatimah, Siti Norasiah Abd Kadir, Affan Nasaruddin, Nobumitsu Sakai Fathiah Mohamed Zuki, Hazreena Hussein, Abdul Halim Sulaiman and Mohammad Shahrul Amin Mohd Salleh. 2018. "Heartware as a Driver for Campus Sustainability: Insights from an Action-oriented Exploratory Case Study." *Journal of Cleaner Production* 196 (September 20): 1086–1096.

Nakamura, M., and W. Rast. 2011. Development of ILBM Platform Process. Evolving Guidelines Through Participatory Improvement. RSCE–Shiga University and ILEC Nicolaides, A. 2006. "The Implementation of Environmental Management Towards Sustainable Universities and Education for Sustainable Development as an Ethical Imperative." *International Journal of Sustainability in Higher Education* 7, no. 4: 414–424.

Nasr, Seyyed Hossein. 1989. *Knowledge of the Sacred*. New York: State University of New York Press.

Suzuki, Ayame. 2014. "The Satoyama Landscapes as a Social-Ecological System: Insights from Harihata and Mukugawa." In *Living Landscapes, Connected Communities: Culture, Environment and Change across Asia*, edited by Justine Vaz and Narumol Aphinives, 70–83. Penang, Malaysia: Areca Books.

Tucker, Mary Evelyn, and John Grim. 2001. "Introduction: The Emerging Alliance of World Religions and Ecology". In "Religion and Ecology: Can the Climate Change." *Daedalus* 130, no. 4: 1–22.

White, Lynn, Jr. 1967. "The Historical Roots of Our Ecological Crisis." *Science* 155, no. 3767: 1203–1207.

Vaz, Justine, and Narumol Aphinives, eds. 2014. *Living Landscapes, Connected Communities: Culture, Environment and Change Across Asia*. Penang, Malaysia: Areca Books.

Chapter 4
Transformative Learning for Thailand's Small-Scale Farmers

Supa Yaimuang

Highlights

- Globalization and monopolies are compromising the sustainability of farming and threatening the incomes of small-scale farmers; Thailand is an illustrative case.
- The Sustainable Agriculture Foundation (Thailand) has worked to create social & economic spaces for farmers, farmer organizations, urban citizens, and other partners in sustainable agriculture development. Three main approaches were used to establish transformative learning: the development of knowledge, the strengthening of farmer organizations, and policy advocacy.
- Farmer awareness of sustainable food production allowed for change and adaptation. Farmers can be innovative both in their farming systems and in their social contexts.
- Issues of sustainability require a regional approach, with the main philosophy being that economic growth must also redress social & economic inequalities, and include improved prosperity of small-scale farmers. Small-scale farmers should participate in policy formulation to strengthen social and economic justice in the region.

The Challenge: Small-Scale Farmers Under Globalization

Thai farmers have faced various changes and challenges from the Green Revolution to the era of high technology and globalization. Under globalization, free-market capitalist ideology drives the global economy. It also dominates patterns of international trade and investment as well as the development of technology used in agricultural production. Advances in information and communications technology have resulted in rapid connectivity throughout the world, spurring the cross-border

S. Yaimuang (✉)
Sustainable Agriculture Foundation, Thailand (SAFT), Bangkok, Thailand

M. Indrawan et al. (eds.), *Civic Engagement in Asia*,
https://doi.org/10.1007/978-981-16-9384-7_4

movement of capital, finance, people, and production sites, as well as news, information, and ideas. All of this has contributed to the growth and increased bargaining power of transnational and large corporations. Trade and investment liberalization have forced domestic agricultural sectors into highly competitive global markets. To keep up, farmers have had to modernize their patterns of production and consumption, impacting their ways of life. Such changes have affected not only farmers, but also the agricultural sector in general. This has had serious implications for the food security and food sovereignty of Thailand.

As the structure of food production and trade has changed, the entire food system has transformed, becoming more and more consolidated and dominated by monopolies. Small-scale farmers find it difficult to participate in the food supply chain. They have limited access to production resources, even genetic material itself. For example, 90% of corn and vegetable seed in Thailand is under the control of just four to five companies (Somporn 2017). This issue is related to intellectual property rights, which have changed the structure of agricultural production and converted common properties into individual properties. In the same way, access to agricultural technology, inputs, and equipment is limited for small-scale farmers. They also find it difficult to access land and water resources.

At the same time, food safety and health problems have become serious issues for both farmers and consumers. A study of the effects of pesticide use on rice farmers in Kalasin province, Thailand, found that 58% of farmers had unsafe levels of chemical residue in their blood and the levels of chemical residue in another 28% of farmers approached unsafe threshold levels (Natawut et al. 2014, 306). Nearly one-third of consumers had unsafe levels of chemical residue in their blood, and over 50% were at risk (Prachachat Life-Health 2015).

While globalization has changed patterns of consumption, easy access to news and information has changed consumer behavior. In Thailand, the consumption of industrially produced fast food has replaced that of local food, particularly among younger generations. Fast-food businesses in Thailand grow nearly 10% per year and in 2016 provided the agri-food industry with a staggering total sales value of 34 billion baht (Marketeer Team 2016).

Such changes in consumption patterns reverberate throughout the agricultural production system. As fewer varieties of raw ingredients are being consumed, the production system, in turn, focuses on cultivating fewer varieties. The new food culture departs in a significant way from the traditional one based on ingredients from local varieties grown by farmers in communities in accordance with the agro-ecosystem of each region. Food has come to be considered more of a commercial commodity than a necessity of life and is consumed more in line with business promotion and advertisements rather than according to human needs. Health problems that are linked to Thailand's new eating habits, such as obesity, diabetes, and heart disease, are also on the rise.

Farmer access to the marketplace has also become limited. The agro-food sector has expanded to include not only upstream agricultural inputs but also downstream retail food outlets. Department stores, supermarkets, food stores, and convenience stores have rapidly replaced fresh markets and corner shops in urban communities in Thailand during the past three decades, particularly after the 1997 financial crisis. With modern technologies and management efficiency, wholesale and retail stores have captured up to 50% of food sales in the country, according to Poolsuk and Piyanuch (2017, 7). In the aftermath of the financial crisis, various food stores were sold to foreign shareholders and foreign direct investment (FDI) in the food and beverage sector increased, helped by the deregulation of FDI. The authors report that food stores expanded into suburbs and the number of supermarket chains increased from just 18 in 1996 to 120 at the end of 2006; in 2010, TESCO alone had 600 branches that altogether bought 1.5 billion *Baht* (approximately 45.2 million USD) worth of domestic agricultural goods (2017, 7). In 2013, there were 12,996 convenience stores in Thailand, but the number increased to 15,325 in 2016 (Poolsuk and Piyanuch 2017, 7).

The Rise of SAFT

The Sustainable Agriculture Foundation, Thailand (SAFT), works with small-scale farmers and city dwellers in Thailand. SAFT promotes sustainable agriculture systems and urban farming. SAFT's vision is to create an agricultural system that is ecologically sound, environmentally friendly, and that contributes to building a just economy and society that supports relationships among humans and between humans and nature.

SAFT was founded in 1998 with the cooperation of the Alternative Agricultural Network. The aim was to create an organization that would work in the knowledge sector and develop academic research in collaboration with farmers and communities, and to connect with other groups of people in society. SAFT's role is that of providing learning support and conducting advocacy campaigns to promote sustainable agriculture and good nutrition. The organization's work strengthens farmers and rural communities in the realm of economic improvement, food production, and environmental protection through developing various sustainable agriculture systems and improving on-farm biodiversity based on the local knowledge systems of farmers and communities. The development of knowledge happens simultaneously with action through participatory research and development, responding to various movements in all areas and at all levels, including producing, processing, marketing, mutual resource management, creating cooperative societies in both urban and rural areas, and making changes at the policy level.

SAFT's academic role is to drive the development of knowledge through focusing on the development of ideas and innovations in sustainable agriculture with small farmer organizations. At the same time, we coordinate with small enterprises, the public sector, and academic institutions. SAFT also supports research, enhances learning processes, and promotes local knowledge in collaboration with farmers and community organizations. It promotes and develops knowledge and innovations in both rural and urban farming, helping to link food systems and alternative lifestyles concerned with nature and creating a society that fosters relationships among urban people as well as relationships between urban and rural people. SAFT develops public media and campaigns about sustainable agriculture, urban farming, and food security and promotes good relations between producers and consumers. SAFT partners with organizations in the public and private sectors, along with civil society, to campaign on policies in sustainable agriculture and related areas. It creates social spaces for farmers, farmers' organizations, urban citizens, and other partners in sustainable agriculture development to advocate for sustainable agriculture to the wider public.

Changing the Learning Processes

SAFT has been implementing solutions based on participatory processes with farmer organizations and networks. Fundamental to this is the learning process. SAFT has a three-pronged strategy to realize alternative development: the development of knowledge, the strengthening of farmer organizations, and policy advocacy.

The first strategy, the development of knowledge and innovation, pertains to sustainable agriculture systems and local seeds. The learning process focuses on raising awareness of small-scale farmers through problem analysis. Analysis of food security in the community and the structure of the food supply chain can strengthen the capacity for analytical thinking not only for farmers but also for city dwellers. This has led to developing knowledge and innovations in organic farming, urban farming, and on-farm biodiversity, and providing organic and safe food products to consumers. Small-scale farmers analyze their economic problems related to monoculture production and investment costs at the household level, based on their own life experiences and knowledge. A timeline is used as a tool to review changes in their own lives and the community. Information on the history and development of Thai agriculture and trade policy under globalization is added to the analysis. Apart from economic problems, environmental and health problems such as biodiversity loss and the impact of chemicals are also analyzed.

Following critical analysis of the problems, solutions are developed; for example, helping small-scale farmers transition away from agro-chemical systems toward sustainable agriculture. Learning processes among small-scale farmers at the household and group levels that identify solutions are systematically developed. Internal transformation processes for farmers and groups, including the process of learning from outside, such as through study trips and discussion among groups, are therefore facilitated. Some workshops and training on techniques are also organized for farmers who want to transform their farming systems with organic or sustainable methods.

Another important learning process is participatory action research and on-farm demonstration. The research may be conducted after returning from a study visit or after problem analysis. The goals and objectives of the research are set by small-scale farmers' groups themselves to solve problems facing individual farmers and the community at large. Research methodologies vary, including interviewing elderly farmers who hold local wisdom, conducting field experiments, and cooperating with

NGOs, academics, and government agencies to collect information. Findings are summarized and analyzed together to build new efficient and sustainable farms. This creates confidence in knowledge-generating processes that lead to changes in attitudes and beliefs about the food system. Before such changes, farmers believed that if they did not use chemicals and pesticides, they would not have good crop yields. After deep study and implementation, however, those beliefs and attitudes have changed. For example, smallholder farmers in one participating village in Yasothon province conducted a comparative study of organic and non-organic farming. The findings revealed that organic yields were low in the beginning but high in the long run, and in the long term, higher intensive production was possible. This contrasted with conventional farming, where yields were high in the beginning, but low in the long run, and lower productivity was the norm in the long term. After conducting the study, farmers adapted their farming methods.

Thailand now produces many organic products, including rice, vegetables, fruits, herbs, tea, and coffee. However, the main product is organic rice. Green Net has reported that the area under organic cultivation in Thailand increased from 235,523 *rai* in 2014 to 284,918 *rai* in 2015 (Green Net 2017), a 21% increase.[1] Furthermore, in the same period, the number of organic certified farms increased from 9,961 farms in 2014 to 13,154 in 2015. The value of sales from Thai organic agricultural products in 2014 totaled 2.3 billion baht (Green Net 2017).

The development of knowledge by farmer groups and communities concerning production systems and genetic conservation has led to increased on-farm biodiversity, demonstrating a concrete solution to the problem of decreasing biodiversity. Small-scale farmers conserve, select, and improve local varieties of rice, fruits, and vegetables, safeguarding genetic resources and enhancing the breeding potential to produce new species, including through in situ practices. In the wake of the Green Revolution, the genetic improvement of agricultural cultivars was perceived as the role of the private sector, government agencies, and academic institutions, while small-scale farmers were largely sidelined. Now, it has been proven that even small-scale farmers can conserve hundreds of varieties of rice, fruits, and vegetables, enabling diversity in our nutritional intake and preserving an important resource base for present and future agricultural systems. When farmers have knowledge and awareness about the seeds they are using, they can play a significant role in conserving existing cultivars and developing new ones.

The second strategy, to strengthen farmer organizations, aims to create a safe food production system. Farmer organizations, through mutual help and sharing, play an important role in building capacity in critical thinking and management, ultimately contributing to determining the direction of development. Farmers organize into groups to manage members' products and marketing, including building relationships with consumers through farmer-owned markets that allow them to directly share information about their production methods. Such communication with consumers creates learning opportunities and mutual understanding within the

[1] 1 *rai* = 6.25 hectares.

process of producing and consuming food. Farmers' markets have changed the relationship between producer and consumer, and the learning process can eventually be extended to many provinces.

Market management also builds the management and sales skills of farmers. In the past, farmers did not have a strong role in this arena, usually selling their products through middlemen. By directly interacting with consumers, small-scale farmers can gain knowledge and understanding about consumer needs. Mutual exchange leads to improvements in the production system, such as providing more varieties of food. Farmers' or "green" markets are a participatory platform in the food supply chain that provides an alternative for farmers and consumers and positively contribute to the economy of farmers' households and communities.

Finally, it is important to link the model of sustainable farming and farmers' marketing to policy. Farmer organizations campaign for policies that support farmers' rights, community rights, and sustainable agricultural systems. This will result in the broader development of a sustainable agricultural system.

The recent market demand for safe food and the concrete examples of farmers and communities changing to organic farming has influenced the adoption of new policies. Organic products are safe for household use and can be exported, contributing to the state economy. Due to these outcomes and the policy campaigns of various organizations, the government finally accepted a sustainable agriculture policy, setting the target of 5 million *rai* of farmland under sustainable agriculture production in the 12th National Social and Economic Plan (2017–2021), commonly referred to as "Plan 12." Plan 12 established committees composed of small-scale farmers, academics, non-governmental organizations, and the government to achieve this production target for sustainable agriculture.

Farmer organizations are campaigning to protect community rights to seed in the law on plant variety protection (PVP) and in opposing Genetically Modified Organisms (GMOs). They also oppose the prioritization of individual rights over community rights when it comes to the issue of Intellectual Property Rights under the International Union for the Protection of New Varieties of Plants 1991 (UPOV 1991) and the agreement of free trade and Intellectual Property Rights under the Regional Comprehensive Economic Partnership (RCEP). These agreements have serious effects on farmers and communities and do not give value to local genetics and community wisdom. In the face of this, farmers and communities are exercising their rights to protect local seeds and wisdom. SAFT and its network have been active in supporting farmers in their policy advocacy efforts.

The learning process to create alternative solutions has led to changes in practices at the field and group management levels. These changes include those in farmer skills, knowledge, and beliefs, as well as in livelihoods. The paradigm of agriculture has been changed into one that integrates economic with health and social dimensions. Raising farmer awareness has led to adaptation and change. Farmers are innovative both in their farming systems and in their social contexts. On the farm, they do not just change from chemical to organic techniques; they implement systemic innovations based on adapting both local and scientific knowledge, old and new technology. For example, farmers have invented small tractors suitable for their farm size and built greenhouses to specialize in particular vegetables. They have collaborated with university academics to analyze soil nutrients and how to most efficiently utilize them in organic farming systems. Farmers save their seeds and create ecologically sound farming systems to maintain food sovereignty. They also innovate in the post-harvest stage, adding value, and marketing. For example, after a scientific study on the nutritional value of local rice varieties, farmers set up a community food-processing group to produce rice desserts. Farmers have established markets that are jointly managed with consumers, building new relationships across the producer–consumer divide. Farmers are also forming groups to help each other in terms of funding, knowledge sharing, and problem-solving. This contributes to social innovation at the community level.

Issues of Sustainability

Sustainable agriculture and small-scale farmers face several challenges. Socio-economic changes due to globalization have impacted the rural economy and society. Climate change effects, such as severe storms or unprecedented temperatures, impact agricultural production and ecosystems and increase natural disaster risks. Small-scale farmers need to innovate and develop knowledge of agricultural techniques and ecological recovery. Sustainable agriculture policy is included in Thailand's National Social and Economic Plan; in practice, however, the strategies, plans, and actions used to encourage farmers to farm sustainably are still decided in a top-down manner. Participation by farmers and communities in policy-making remains weak, and they cannot access support from the government.

Several important policies affect the agricultural sector, including those on special economic zones, land use, power plant development, and mining development. The dominant development paradigm is linked with the borderless economic framework in the region. The driving forces behind policy-making focus on the economy rather than on farmers and community livelihoods or the environment, even though ecological and environmental changes—that are borderless—are beginning to pose a threat to the agricultural sector. Regional cooperation is key to positive socio-economic change. Economic growth in the region must include improved prosperity of farmers and local communities in order to be sustainable.

Collaboration among small-scale farmers and networks across the country through the exchange of knowledge and sharing of practices is necessary to strengthen the economic, social, cultural, and environmental standing of farmers and communities. Consequently, ASEAN members should encourage small-scale farmers to participate in policy formulation intended to strengthen social and economic justice in the region.

References

Green Net. 2017. "Overview of the Situation of Thai Organic Farming in 2016." https://www.greennet.or.th/th-organic-2559/.

Marketeer Team. 2016. "Talat Ranahan Fast Food" (Fast Food Market Share). Marketeer Online website. https://marketeeronline.co/archives/24956.

Natawut Paipard, Somjit Supannatas, and Teerapat Suttiprapa. 2014. "The Effects of Pesticide Use on Rice Farmer's Health and on the Environment in Rong Kham District, Kalasin Province" (Thai language source). *Khon Kaen Agricultural Journal* 42, no. 3: 301–310. https://ag2.kku.ac.th/kaj/PDF.cfm?filename=05%20Natawut.pdf&id=1711&keeptrack=2.

Poolsuk Nilkijsaranont, and Piyanuch Stapongpukdee. 2017. *Thurakit Rankha Plik Samai Mai* (The Modern Retail Trading Business). Bangkok: Bank of Ayudhya Public Company Ltd. https://www.krungsri.com/bank/getmedia/683cacb7-e02f-4bee-a3cc-17827b26c929/IO_Modern_Trade_2017_TH.aspx.

Prachachat Life-Health. 2015. "Khon Thai Siang To Sankhemi Tok Khang – Nae Chai Nam Som Sai Chu Lang Phak Phonlamai" (Thai People are at Risk of Chemical Residues—Are Advised to Use Distilled White Vinegar to Wash Vegetables and Fruit). https://www.prachachat.net/news_detail.php?newsid=1434368613.

Somporn Isvilanont. 2017. "Utsahakam Maletphan khaung Thai: Sathanaphap lae Khwamthathai" (Thai Seed Industry: Status and Challenge). Lecture given at the 14th Academic Conference on Seeds of Thailand, at the King Mongkut's Institute of Technology Ladkrabang, November 30. http://seed.or.th/documents/thseedindustry.pdf.

Chapter 5
The Role of Citizen Science in Policy Advocacy and Building Just and Ecologically Sustainable Communities in Thailand

Penchom Saetang

Highlights

- Citizen science can be a powerful tool and approach to support the building of a just and ecologically sustainable society in Thailand.
- Empowering participatory roles in dealing with state authorities and citizen capacity for monitoring polluting corporations can increase good governance and transparency in industrial pollution management.

Introduction

Since the early 1980s, the Thai government has focused on industrializing its economy. Investment in the industrial sector has been heavily promoted and has rapidly increased without corresponding environmental regulations and enforcement. Pollution management has often included the use of "end of pipe" technologies only, merely requiring the treatment or filtration of a few pollutants before effluents are discharged into the environment. For decades, industrial factories in Thailand have operated without official regulations, stringent technological requirements, or post-licensing auditing to minimize pollution emissions. As such, each year, factories generate millions of tons of hazardous waste and emit huge volumes of air and water pollutants. These toxic substances accumulate over time and are often invisible; it is, therefore, difficult to estimate their true harm and cost to human health and the environment.

Presently, the Thai government has committed itself to achieving the Sustainable Development Goals (SDGs) set out by the United Nations. Despite this commitment, the government's promotion of large-scale investment and environmentally-unfriendly economic development continues unabated. Indeed, several recently

P. Saetang (✉)
Ecological Alert and Recovery—Thailand (EARTH), Nonthaburi, Thailand

M. Indrawan et al. (eds.), *Civic Engagement in Asia*,
https://doi.org/10.1007/978-981-16-9384-7_5

approved laws, regulations, and orders are designed to further attract industrial investment by providing special services and privileges to investors. These include the Investment Promotion Act, B.E. 2560 (2017) and the Competitiveness Enhancement for Target Industries Act B.E. 2560 (2017), both of which will establish a one-stop service center for the Eastern Economic Corridor (EEC).[1] At the same time, very little has been done to improve environmental quality. Inadequate enforcement of the existing laws, to dissuade polluters from aggressive exploitation, has continued. The "end of pipe" approach to managing pollution has not been substituted with the best environmental practices and best available technology (BEP/BAT) approach. Such a transition has never even been put under serious consideration or put to public debate. Consequently, citizen demonstrations and petitions against the damaging impacts of pollution have endured and increased over time. Many communities affected by pollution have been organizing to protect themselves from environmental risk and to try to stop more destruction of nature. Such activism has emerged in various places across the country during the past several decades and is critical to supporting and building a sustainable society for our future.

This chapter describes how industrial pollution in Thailand has compromised public safety and how inadequate government and industry response has given rise to citizen science. A public engagement approach to scientific research, citizen science has served as a guide and working method developed through cumulative experiences in pollution campaigns and with the support of environmental experts from inside and outside the country. In many countries around the world, citizen science is playing an important and increasing role in protecting the environment, conserving endangered species, and protecting human health from harmful pollution[2] (Kullenberg and Kasperowski 2016).

Chronic Conflict Caused by Polluting Industries

Environmental problems related to industrial pollution have been evident in Thailand since the late 1990s. The widespread and grave impacts of pollution on waterways, farmlands, air, human health, and livelihood have pushed communities to demonstrate on the streets. Some cases have turned into deep conflicts between aggrieved peoples and the authorities and industrial corporations; several have become controversial issues that have remained unsolved for decades. One of the most well-known cases is that of the Phong River in Khon Kaen province of Thailand's northeastern region. After several years of disputes between local communities and factories in Nam Phong district, the situation became intolerable in 1997, when more than 500,000 fish raised in floating baskets died due to suspected effluents from the

[1] Incentives under the Investment Promotion Act. https://www.boi.go.th/index.php?page=incentive (Accessed 9 December 2019).

[2] Takagi Fund for Citizen Science. http://www.takagifund.org/e/about/citizenscience.html (Accessed 9 December 2019).

Phoenix Pulp and Paper Company. Huge numbers of dead shellfish and aquatic fauna were also found in the Phong River and its tributaries (Inmuang 2018). Local villagers, who relied on fishing for their livelihood, faced immediate bankruptcy. As it was difficult to prove the Phoenix Pulp and Paper Public Company was responsible for the loss, the affected people had to demonstrate in the streets to pressure for compensation. The company finally compensated communities with 2.9 million baht (approximately 92,000 USD) for the fish deaths; however, the polluted Phong River has remained an unsolved problem to this day.

Mae Moh

In Mae Moh district, Lampang province, people in 16 villages of 4 subdistricts have fallen into a long dispute with the Electricity Generation Authority of Thailand (EGAT), after it started operating its first coal-fired power plant in 1978. Today, there are 13 coal-fired power plants in Mae Moh. The latest, with a total capacity of 2,625 megawatts, began operations in 1995 (EGAT 2016). Air pollution from power plants has caused extensive damage to local farmlands, crops, and livestock. Villagers' health and livelihood have also been badly affected. Large numbers of people have fallen ill and increased rates of premature death have been recorded. From January to June 1998 alone, the Lampang Provincial Health Office reported that 3,463 (out of 8,214 people who received medical examinations from the Provincial Health Office) were found sick with *respiratory system diseases* (Rojanapaiwong 2000, 294). Between 1996 and 2005, the central government and EGAT set up at least ten committees and working groups consisting of officials, health experts, and concerned parties to solve problems caused by lignite mining and coal-fired power plants. During that period, several thousand villagers were treated for respiratory disease symptoms.

Regular monitoring of air pollution and health impacts has been conducted in many villages. This helped some villages receive compensation and led to the relocation of others (Srisawat 2005, 19–24). Compensation, however, was very small compared to the extensive and chronic impacts suffered by more than 10,000 villagers in Mae Moh district. During the last decade, several hundred affected villagers have sued EGAT for compensation for health and environmental damages. In 2015, the Supreme Administrative Court based in Chiang Mai province directed EGAT to pay 49 million Baht (approximately 1,538,500 USD) with interest to compensate 123 Mae Moh villagers; according to the court order of March 13, 2015, EGAT paid the damages (EGAT 2015).

At present, although there have been some improvements to air quality, for example through the installation of the Flue Gas Desulfurization System (FGD), the application of technologies to improve operational efficiency in managing dust and contaminated water, reductions in noise pollution, and new steps to prevent coal ignition (EGAT 2019), severe pollution persists. Learning from the painful lessons of Mae Moh, many communities in other provinces have spearheaded significant opposition and protests against new coal-fired power plants previously planned for Prachuap Khiri Khan, Songkhla, Krabi, Rayong, and Chachoengsao. The experiences of Mae Moh, as one of the worst pollution hotspots in the country, continue to reverberate.

Map Ta Phut

The Map Ta Phut area in Rayong province is the largest petrochemical industrial complex in Thailand. Before the Eastern Seaboard Development (ESB) Program began in 1987, this area was considered one of the best and most well-known touristic beaches near Bangkok, due to the coexistence of beautiful sand beaches and coastal biodiversity, rice fields, fruit orchards, economic plantations, and forests. The development of the industrial deep seaport led to the destruction of over 10 miles of coastline. Additionally, large areas of paddy fields, orchards, and living areas have been expropriated since 1985 and converted into large industrial estates that house almost 100 large factories. At the beginning stage of industrial development, Map Ta Phut villagers happily welcomed the changes and waited to reap the benefits of economic development and employment in their hometown. In actuality, the industrial growth has engulfed the villagers, schoolteachers, students, and resort operators in all kinds of pollution, endless construction of factories, and round-the-clock transportation of goods and hazardous substances. Several thousand have suffered from air pollution emitted from major industries, including petrochemical plants, oil refineries, coal-fired power stations, and iron and steel facilities.

Villagers tried to address these problems and submitted several petitions about the health and environmental impacts, but nothing improved. In 2007, people from eleven communities in the Map Ta Phut zone filed a lawsuit against the National Environment Board (NEB), alleging that the board had improperly failed to designate Map Ta Phut and its vicinity a pollution control zone (EJAtlas 2018). NEB finally issued the pollution control zone order in April 2009.[3] In September of that year, the Administrative Court decided on another lawsuit, which was filed against eight responsible agencies, including the NEB, to suspend 76 new petrochemicals and related industry projects in Map Ta Phut (Administrative Court 2009). Despite the decision, 74 of the 76 contested projects were allowed to continue operations in 2010 (Reuters 2010).

The litigation movement has pressured the Thai government to find more sustainable ways to achieve industrial growth in Map Ta Phut. Consequently, the Japan International Cooperation Agency (JICA) recommended that the Thai government run a trial model of the Pollutant Release and Transfer Register (PRTR), a database system that would provide public access to information on the amounts of pollutants released from different point and non-point sources of water, air, and soil.

[3] National Environment Board Announcement No. 32 (BE 2552), 30th of April 2009 (Thai language source).

This included data on the solid waste that was transferred off-site for disposal and wastewater treatment. The PRTR system, which is mandatory in many industrialized countries, such as the United States, European Union countries, and Japan, has incentivized industry and businesses to reduce pollutant releases and transfers of waste as well as to adopt cleaner production techniques that successfully improve environmental quality.

In July 2010, three Thai agencies, the Pollution Control Department (PCD), the Department of Industrial Works (DIW), and the Industrial Estate Authority of Thailand (IEAT), together with JICA as the technical consultant, began a technical cooperation project for "The Development of Basic Schemes for Pollutant Release and Transfer Register System (PRTR) in the Kingdom of Thailand," with the main objective of developing a model for a pilot PRTR system in Thailand. Rayong province was chosen to launch the pilot scheme since Map Ta Phut and areas nearby were known as the major source of industrial pollution. Under this project, a total of 1,369 industrial facilities (point source polluters) in Rayong were required to voluntarily submit estimation reports of 107 substances emitted into the environment (air, water, and soil). Of these, 525 facilities housed in industrial estates were asked to voluntarily submit emission reports to the IEAT and the other 844 located outside industrial estates were asked to voluntarily submit reports to the DIW. Meanwhile, the PCD would be responsible for estimating the annual emissions of the transportation and agricultural sectors and waste management facilities (PCD et al. 2016).

Despite the potential of the database system, seven years after its implementation, the World Resources Institute denounced Map Ta Phut as "one of Thailand's most toxic hotspots with a well-documented history of air and water pollution, industrial accidents, illegal hazardous waste dumping, and pollution-related health impacts including cancer and birth deformities" (World Resources Institute 2017).

Industrial Expansion Without Pollution Solutions

In June 2016, the Thai government approved the Eastern Economic Corridor (EEC) program, despite predictions that the anticipated environmental and health-related problems of the program would be severe and unsolvable. The EEC is a new phase of industrial expansion covering Rayong, Chonburi, and Chachoengsao provinces. The expansion includes a supercluster of petrochemical and chemical industries, new deep seaports, and the Smart Park project, a new large industrial area that will promote hi-tech industries such as automotive, smart electronics, and biotechnology. The approval prompted serious concerns within nearby communities of increased conflicts due to more land being expropriated and the industries causing more environmental and health problems (Saetang and Thassanaphak 2017, 156–157). The fishing communities expressed their worries about deterioration of seawater quality, while the construction of the new industrial deep seaport would change the water currents and ruin longer stretches of coastal areas than what had already been destroyed. Environmentalists also admonished that the EEC would intensify conflicts with and within local communities and aggravate existing environmental destruction while providing no compensation to pollution victims (Saetang and Mooksuwan 2003, 102). The program, however, is going ahead. According to statistics of the Board of Investment (BOI), in 2017 there were direct investments in the EEC region of 259 projects valued at 310.337 billion THB. The projects and their values were listed by province as follows: Chonburi—133 projects valued at 117.311 billion THB; Rayong—93 projects valued at 162.751 billion THB; and Chachoengsao—33 projects valued at 30.275 billion THB (EEC 2018).

Industrial pollution in Thailand is now widespread and its consequences are evidenced by a loss of vegetation and biological diversity and excessive amounts of harmful chemicals in the ambient atmosphere and food chains (Sathirathai et al. 2002). Polluters have rarely been held accountable due to a lack of scientific evidence to firmly pinpoint the causal relationship, particularly when several polluters are situated in the same estate. For example, to identify the air polluters at the Map Ta Phut Industrial Estate (MTP IE) between 1997 and 1998, local authorities and the PCD set up a "sniffing team," which was tasked with identifying the sources and causes of chemical smells that made hundreds of school students and villagers fall ill. The team consisted of concerned parties, including the affected schoolteachers and students. This simple and grossly inadequate method was applied to solve the problems because no factories accepted fault, while the authorities claimed that no efficient tools were available for them to detect the source of the chemicals. The sniffing team was finally able to identify six out of eighty-nine factories in the Map Ta Phut Industrial Estate that emitted a foul smell. These six factories, most of which were involved with oil refining and petrochemicals, were ordered by the IEAT, the regulatory agency, to stop operation until they were able to fix the air pollution problems and install additional systems to reduce chemical emission. The inadequacy of this approach is further highlighted when one considers that many odorless chemicals are dangerous to human health. Such inadequate response and lack of accountability are common across Thailand's industrial sector.[4]

Challenges in Regulatory Enforcement

The Enhancement and Conservation of National Environmental Quality Act, B.E. 2535 (1992), Thailand's basic environmental law, adopts the Polluter Pays Principle (PPP) in its pollution control chapter. In practice, however, the rules outlined in this chapter have rarely forced polluters to be responsible for damages to the environment

[4] For example, industries have not been held responsible for the widespread impacts of sulfur dioxide in Mae Moh District; poisoning of the Phong River; lead contamination in Lower Klity Village in Kanchanaburi province; gold mining and arsenic contamination in Wangsaphung District, Loei Province; methylmercury accumulation in fish and human hair in the Tha Tum area or Prachinburi Province; and volatile organic compounds (VOCs) in groundwater in the Tha Tan area of Chachoengsao Province.

or the health of people. This is due to difficulties in proving their guilt in court, particularly in cases where scientific evidence is needed to directly link toxic emissions from factories to illnesses and destruction of the environment. Besides, ministerial regulations and announcements issued following the pollution control chapter cover only some specific emission standards for designated pollution sources (Rowbotham et al. 2017), while many types of toxic chemicals still lack such standards to set quantitative limits on permissible emission amounts.

Another obstacle to holding polluters to account is that emission data and related environmental auditing reports are held solely by government agencies; they are not accessible to the public. For this reason, the principles of the right to know and of public participation have never been pragmatically applied in the country, despite the Thai government's endorsement of the Rio Declaration of 1992 and ratification of other international conventions that strongly promote such principles. Consequently, to date, only a small number of pollution cases have won in civil court and a very small percentage of affected peoples have received moderate compensation. In each case, before successful litigation, the claimants had to tolerate years of suffering, applying many social and political methods to draw the attention of the general public, the media, and the government.

Citizen Science in Thailand

In locations where pollution problems are severe and residents have inadequate information, environmental monitoring by communities is a practical way to collect and record data about actual pollution and its impacts. This method can create an impressive and reliable body of evidence that is useful in mobilizing environmental campaigns, negotiating with the polluters, and testifying in a court of law. EARTH (Ecological Alert and Recovery – Thailand) established a citizen science project in 2015 to support community monitoring. It believes that communities with environmental monitoring training have sufficient capacity to generate strategic evidence for their fight against pollution problems.

EARTH's first experience using a citizen science method was to prove and monitor toxic contamination in the ambient air around the Map Ta Phut Industrial Estate during 2004–2005.[5] In collaboration with Greenpeace Southeast Asia (GP SEA) and with technical support from Global Community Monitor (GCM) based in the United States, EARTH collected five air samples for chemical analysis. Lab analysis identified a total of 20 different toxic chemicals in the five samples. A minimum of six and a maximum of twelve volatile organic compounds (VOCs) and sulfur compounds were detected in each sample, and at least two of the toxic chemicals in each sample exceeded one or more health-protective standards or screening levels.

[5] In the period 2004–2005, EARTH was known by its former name: Campaign for Alternative Industry Network (CAIN). The name was changed to Ecological Alert and Recovery - Thailand (EARTH) when it was registered as a foundation in 2009.

This provided concrete evidence of the toxic cocktail inhaled daily by Map Ta Phut residents.

One of the most alarming VOCs, detected in four of the five samples, was benzene, known as a human cancer-causing agent. The amounts of benzene present exceeded the United States' Environmental Protection Agency's Annual Ambient Air Screening Level by as much as 60 times. Several other cancer-causing agents were discovered to exceed screening levels by as much as 3,378 times.[6] The analysis results were published by campaign groups and used in a press conference to address the need for the designation of VOC screening levels in Thailand. EARTH also drew upon the results to argue that regular monitoring of VOCs is urgently needed to protect the health of Map Ta Phut residents from these harmful substances. The findings, though insufficient to prove the relation of the toxic chemicals and cancer diseases developed in many Map Ta Phut residents, were significant in convincing the National Environment Board to issue an Announcement on the Annual Ambient Air Screening Level of 9 Volatile Organic Compounds in 2007, and in 2009, the PCD issued another announcement about VOCs regarding a 24–h Ambient Air Screening Level of 19 Volatile Organic Compounds. Subsequently, the PCD set up VOC monitoring stations to regularly report VOC levels in air and groundwater in the Map Ta Phut area.

In 2012, EARTH further developed the citizen science approach to support pollution-affected communities. It received a citizen science research grant program from the Takagi Fund to conduct collaborative research with affected communities close to a gold mining area in Loei province to detect chemical contamination in the environment. The environmental samples included three soil samples from the villagers' farming areas, four sediment samples from local waterways, eight surface water samples, three samples of rice plants, three samples of paddy seeds, and four tap water samples from groundwater sources. Arsenic was the heavy metal most commonly found in all types of samples, but in varying levels. One of the soil samples found arsenic levels 17 times higher than the standard level of soil quality for agricultural purposes (the standard level is 3.9 mg/kg). Some other substances, such as mercury and lead, were also detected in the soil, but at minimal levels. All surface water samples registered arsenic levels higher than the water quality standards for surface water as designated by the NEB, and all of the four sediment samples were found to have arsenic levels higher than the toxic substance screening level in sediment of the EPA (the US Environmental Protection Agency), which is 9.8 mg/kg (at the time, no heavy metal standard level existed in Thailand). Lastly, three samples of rice plants and three samples of rice seeds were also found to be contaminated with arsenic and lead in varying levels, but no concentration standards for heavy metals in paddy plants and seeds were available for comparison. The finding that rice seed and drinking water were contaminated with arsenic was striking to the villagers and

[6] The amount of vinyl chloride detected in two samples exceeded the EPA screening level by as much as 86 times; the amount of 1,2-Dichloroethane (EDC) was present in two samples and exceeded the EPA screening level by as much as 3,378 times; chloroform, also known to be a probable human cancer-causing agent, in one sample was in excess of the EPA Annual Ambient Air Screening Level by 119 times (Buakamsri et al. 2005).

prompted them to campaign to stop the second phase expansion of the gold mines. At the same time, the affected villagers joined a health impact survey led by a group of academics from the Loei Provincial Health Office and the National Health Commission Office to investigate arsenic levels in villagers' bodies. They proposed a petition to the PCD and the Provincial Governor to consider developing a remedial plan in the contaminated areas and finally filed a civil lawsuit at the Loei Provincial Court to claim for damages to health and household economy in 2018.

From 2015 to the present, EARTH has carried out citizen science projects around the country to increase the negotiating power of communities affected by industrial pollution in their demands for corporate and government accountability. To date, we have provided advice to around 40 communities in 15 provinces in their struggle to protect the environment. We have also helped three communities apply the citizen science method in monitoring and collecting data and environmental evidence for legal battles. The Khao Luang Heritage Conservation Group in Loei, for instance, implemented the Gold Mining Pollution Monitor Project with EARTH's support to strengthen their environmental surveillance and develop a community learning center for health and environment. Field research and data collection during 2015–2018 has significantly supported their victory in the Provincial Civil Court's decision on December 13, 2018, which ordered Thungkham Co. Ltd., the gold mining company in Loei province, to remedy contaminated areas both inside and outside the mining areas and pay compensation of 104,000 Thai Baht with 7.5% interest (approximately 3,265 USD) per family to 149 families.

Concluding Remarks

EARTH has developed the knowledge and technical skills to strengthen environmental safeguarding by local people. We believe this will be beneficial not only to communities in contaminated environments, but also to local and national government agencies, civil society groups in Thailand and some neighboring countries, academics, and, most importantly, the industrial sector, which will benefit from increased dialogue with local communities. Ultimately, it is an approach that can support the creation of a just and ecologically sustainable future for Thailand and other countries that are facing similar environmental threats (Figs. 5.1, 5.2, 5.3, 5.4, 5.5, 5.6 and 5.7).

Fig. 5.1 EARTH's first experience using a citizen science method was to prove and monitor toxic contamination in the ambient air around the Map Ta Phut Industrial Estate between 2004–2005, in collaboration with Greenpeace Southeast Asia based in Bangkok, and the technical support of the Global Community Monitor (GCM) based in the United States. Five air samples were collected for chemical analysis using a scientific tool called "the Bucket." The "Bucket" is a sturdy, easy-to-use plastic container with a special sampling 'Tedlar' bag attached to capture gases. It was created in 1995 and has been verified as a credible method for air sampling by the US Environmental Protection Agency (EPA). The five samples found a total of 20 different volatile organic compounds, including carcinogens

Fig. 5.2 EARTH and the experts from the Arnika Association from the Czech Republic conducted a collaborative study about the heavy metal contamination in sediment in eight pollution hotspots of the country, under the project "Increasing Transparency in Industrial Pollution Management through Citizen Science 2015–2018," funded by the European Union and the Thai Health Promotion Foundation. The pictures show the collection of sediment samples from a waterway near the pulp and paper and coal-fired power plant in Nam Phong district, Khon Kaen province

Fig. 5.3 The author and villagers in Tha Tum subdistrict, Prachinburi province discuss the specific location to conduct sediment sampling in February 2016, under the project "Increasing Transparency in Industrial Pollution Management through Citizen Science 2015–2018" project

Fig. 5.4 Khao Luang Heritage Conservation Group, a self-organized group of villagers affected by gold mining in Loei province. They have implemented the Gold Mining Pollution Monitor Project with EARTH's support to strengthen environmental surveillance and develop a community learning center for health and environment

Fig. 5.5 EARTH (Penchom Saetang) and Arnika Association (Jindrich Petrlik) held a press conference in November 2017 to release study results about dioxin accumulation in free-range chicken egg samples collected near recycling areas in Samut Sakhon province, part of the project "Increasing Transparency in Industrial Pollution Management through Citizen Science 2015–2018."

Fig. 5.6 In early 2016, EARTH conducted a field survey with local community members in Prathum Thani province due to concerns about water source contamination by a proposed waste to energy project near the provincial water source used for tap water production

Fig. 5.7 EARTH collecting water samples close to the municipal waste dump site of Krabi Municipality, Krabi province. The local community affected by improper management of the municipal landfill needed environmental evidence to support their petition for landfill improvement by the municipality

References

Administrative Court. 2009. "Central Administrative Court Issues Temporary Relief Order to Suspend 76 Projects to Protect the Map Ta Phut Communities." *Administrative Court News*. 59/2552. September 30, 2009.

Buakamsri, T, D. Larson, F. Harnnarong, P. Saetang, and W. Mooksuwan. 2005. "Thailand's Air: Poison Cocktail Exposing Unsustainable Industries and the Case for Community Right to Know and Prevention." Greenpeace Southeast Asia, Global Community Monitor, Thailand.

EEC. 2018. "EEC Investment Statistics of 2017." https://www.eeco.or.th/en/content/investment-statistics

EGAT. 2016. "History and Background" (of Mae Moh Power Plants and Mae Moh Mining). Electricity Generating Authority of Thailand (EGAT) website (Thai language source) http://maemoh.egat.com/index.php?option=com_content&view=article&id=71&Itemid=476

EGAT. 2015. "EGAT to pay compensation to Mae Moh villagers on March 13, 2015." https://www.egat.co.th/en/news-announcement/news-release/egat-to-pay-compensation-to-mae-moh-villagers-on-march-13-2015

EGAT. 2019. "EGAT Mae Moh Mine wins Thailand Green and Smart Mining Awards 2019." https://www.egat.co.th/en/news-announcement/news-release/egat-mae-moh-mine-wins-thailand-green-and-smart-mining-awards-2019

EJAtlas (Environmental Justice Atlas). 2018. "Map Ta Phut Industrial Area, Thailand." https://ejatlas.org/conflict/maptaphut-conflict

Inmuang, Y. 2018. "Thailand Water Pollution Crisis: A Case of Massive Fish Deaths in The Phong River, Fact Sheet—Environmental Health." *Department of Health* 1 (9): 1.

Kullenberg, C., and D. Kasperowski. 2016. "What is Citizen Science?—A Scientometric Meta-Analysis." *PLoS ONE* 11 (1): e0147152. pp. 1/16–16/16. https://doi.org/10.1371/journal.pone.0147152.

PCD, DIW, IEAT, and JICA. 2016. "Document on the Pollutant Release and Transfer Register Follow-Up under the Project for the Development of Basic Schemes for the PRTR System in the Kingdom of Thailand, through the Cooperation of the Pollution Control Department (PCD), Department of Industrial Works (DIW), Industrial Estate Authority of Thailand (IEAT), and Japan International Cooperation Agency (JICA)." (Thai language source)

Reuters. 2010. "Update 1-Thai Court Lifts Ban on 74 Industrial Projects." Reuters. September 2, 2010. https://www.reuters.com/article/thailand-industrial-idAFSGE68118P20100902

Rojanapaiwong, S. (ed). 2000. "State of the Thai environment 1997–1998." Green World Foundation.

Rowbotham, S., M. McKinnon, J. Leach, R. Lamberts, and P. Hawe. 2017. "Does citizen science have the capacity to transform population health science?" *Critical Public Health*. https://doi.org/10.1080/09581596.2017.1395393.

Saetang, Penchom and Walaiporn Mooksuwan. 2003. "The Health Impact Assessment of Map Ta Phut Industrial Development and Its Vicinity." Health Research Institute (HSRI) website (Thai language source) https://kb.hsri.or.th/dspace/handle/11228/1219

Saetang, Penchom, and K. Thassanaphak (eds). 2017. State of Industrial Pollution, 2015–2016. Nonthaburi: Ecological Alert and Recovery—Thailand (EARTH). (Thai language source)

Sathirathai, S. et al. 2002. "White Paper on Industrial Development in the Existing Context of Globalization and Its Impacts on Sustainability." World Summit on Sustainable Development 2002, Johannesburg, South Africa.

Srisawat, SC. 2005. "Research Project on Governance and Operation Regulating Mechanisms: the Case of Lignite Mining at Mae Moh, Lampang Province." Independent research paper (Thai language source).

World Resources Institute. 2017. "Thirsting for Justice: Transparency and Poor People's Struggle for Clean Water in Indonesia, Mongolia, and Thailand." https://www.wri.org/publication/thirsting-for-justice

Chapter 6
Creating an Enabling Environment for Lao Youth to Engage with the Community Development Process

Khamphoui Saythalat

Highlights

- With adequate education and advocacy, civic groups may eventually be seen by all stakeholders as valuable, untapped resources rather than as a problematic representation of marginalized communities.
- Youth are central to community development, and therefore long-term commitments to and investments in them need to be made.
- Youth must have platforms and safe spaces for learning and exchange of experiences and knowledge, especially to leverage self-esteem and critical thinking for positive development.

Introduction

Laos has rich natural resources. Yet, significant contributions to the country's socio-economic development have relied heavily on foreign direct investment, with the Lao government introducing several policies and incentives to promote large-scale infrastructure projects, such as dams, mining, mono-crop plantations, and so on. While many good initiatives and projects are undertaken by the government, they oftentimes fail to reach the anticipated goals, particularly due to neglecting the interests of local communities and indigenous groups. The government's development decisions are rarely open to public forums. Effective engagement in the development process, especially one that involves young people aged below 20 years (who represent over 55 percent of the country's total population), is lacking. Surely though, their voices and concerns must be placed on the table of policymakers to ensure Laos's sustainable future.

K. Saythalat (✉)
Participatory Development Training Center, Vientiane, Laos

M. Indrawan et al. (eds.), *Civic Engagement in Asia*,
https://doi.org/10.1007/978-981-16-9384-7_6

Rural communities are especially important. Even today, Laotian rural communities have so far gained only limited benefits from the country's economic and social development. Many of the joint investments for mega-development projects have contributed little to poverty reduction and have, on the contrary, led to increased market externalities. It has become clear to many development partners that living conditions, particularly in rural communities, will not improve unless members of those communities are more deeply involved throughout the development process, from the very early planning stages.

In many parts of Laos, the local government is the key driver of development processes. There is a need to get alternative views of development onto the poverty agenda, convincing authorities of the benefits to be gained by giving more opportunity to local communities to determine their future and allowing for diversity instead of imposing solutions that are often locally unsuitable.

The local governments and communities that are the target beneficiaries of development have yet to understand that promising initiatives such as the National Growth and Poverty Eradication Support Program (NGPES), which included Laos's 47 poorest districts, cannot be sustainable if they are implemented in a business-as-usual manner. For example, since 2016, five villages in Mahaxay district of Khammaoune province (Na Satung, KouanLoung, KhamPeh-Gnai, NaMakBa, and ThongKoung) have not been allowed to use village forests as they traditionally have done to harvest materials for household needs. They are now said to violate Decision 0719 of the Ministry of Industry and Commerce, which was issued in September 2009 (European Commission 2009), and Prime Minister Order Number-15 issued in May 2016 (Gerin and Souksavahn 2016). While the decrees may have intended to produce positive environmental outcomes, by neglecting local community interests they have created new hardships and potential conflicts.

Traditional livelihoods are also threatened. For example, rural Laotians have expressed concerns about diminishing local knowledge, especially weaving activity, as ninety percent of youth today are no longer interested in weaving. As more and more rural young people migrate to find jobs in big cities like Vientiane, Savannakhet, and Pakse, they often fall into a new cycle of poverty that can, in turn, lead to getting caught up in human trafficking, the sex trade, and drug abuse. In 2016, the Lao Ministry of Labor and Social Welfare reported that during 2011–2015 an estimated 781 Lao trafficking victims (75 male) were repatriated from Thailand (Schloenhardt 2017, 60).

Civil society, including community-based organizations, has a key role to play in helping the Lao government fulfill various national and international commitments and obligations, to move smoothly toward good governance and poverty reduction.

Central to the success of civic engagement is the empowerment of youth. Youth need to be able to contribute to society and influence decisions, including those related to local livelihoods. Participatory techniques and learning are indispensable in facilitating young people's civic engagement.

Key Strategies and Methods for Civic Engagement

The Participatory Development Training Center (PADETC) is a Lao non-profit organization established in 1980 as a community development organization promoting sustainable agriculture. In 1996, under the auspices of the Ministry of Education's Department of Private Education, it grew to be an independent training center. The organization uses a participatory development training model in helping Lao people solve their problems in ways that are economically, socially, and environmentally sustainable and just.

Solutions must be formulated in a participatory manner to be sustainable. PADETC firmly believes that *true* engagement with communities must include both technical knowledge and the ability to listen and communicate effectively. PADETC also believes that collective engagement is manifested by equal partnerships and that local civil society organizations and local government agencies may come together to empower a whole community. Hence, our approach is to bring about a participatory consultation process involving key stakeholders (government, communities, and civil society) in which needs can be identified and addressed by the communities with close support from local government agencies, such as district and village authorities. This process ensures participation before signing MoUs[1] or starting projects. A series of consultation meetings are organized, with presentations of the project concept, followed by various discussions and consultations about the roles and responsibilities of each stakeholder vis-à-vis the project implementation.

In its mission to foster civic engagement, PADETC facilitated a multi-stakeholder intervention project called "Capacity Development for Citizen-led Inclusive Development (CD-CID)." The intervention aimed to contribute to inclusive local governance and community development led by the local people, using innovative institutional development approaches in village development planning. Together with development partners,[2] we applied a training tool known as "development facilitation." Through this initiative (2016–2018), 25 villages in four provinces, namely Champasack, Savannakhet, Vientiane capital, and Xiengkhouang, were supported. The curriculum was designed to contribute to transparent, inclusive, and demand-driven community development. The tool involved four key components: (1) consensus to increase and facilitate agreements about the need to enhance capacity; (2) strengthening capacity, including individual competencies and organizational processes, structure, and systems; (3) application of capacities, improved implementation of processes; and (4) building of structures and systems to improve performance. Next, we undertook monitoring and evaluation using a qualitative approach

---

[1] These MoUs (Memorandum of Understandings) are made to open space for collaboration between PADETC and provincial governments, who together carry out educational activities in cooperation with provincial education offices, namely in Champasak, Savannakhet, Khammouane, and Xiengkhouang.

[2] Oxfam Laos, Cord UK, HMP (Maeying Huamjai Phattana – Lao Knowledge and Information System, Learning House for Development Association, and Vientiane SubCAW) (Sub-Commission for the Advancement of Women), with support from the European Union.

known as the most significant change, or MSC (Sigsgaard 2002). MSC was represented by a visualized story of how organizations in the field of development were enabled to learn from and communicate about the outcomes and impact of their interventions.[3]

PADETC has also successfully applied a community-led empowerment method called Gender in Action Learning System (GALS). GALS is a gender mainstreaming tool developed by Oxfam to help develop ideas and dialogues prior to action. It consists of (1) a set of principles related to gender justice, participation, and leadership; (2) a series of visual diagramming tools that are used for envisioning, analyzing, planning, and tracking change by individuals and households; (3) peer learning and structure for action learning in communities; and (4) a mechanism to integrate GALS into organizations, such as in its financial services and business development. In supporting youth development, PADETC facilitated youth engagement in community activities for livelihood improvement. This is part of a 4-month intensive training course supported by Oxfam Laos (January 2016–December 2018). Thus, in following up on the village participatory research conducted by youth, four projects (with grant awards ranging from $1,000 USD to $4,000 USD) are implemented in the villages, namely:

- Operation of community radio as a means to access information in Ban ThanXiengNuean, Phaxay district, Xiengkhouang province.
- Research of impacts of pesticide use on community livelihoods in Ban Xang, Khoun district, Xiengkhouang province.
- Promotion of organic fertilizer use for household gardens in Ban MeignNaLam, Phek district, Xiengkhouang province.
- Waste management in the rural development agency of Ban Na-Hoi, Sangthong district, Vientiane capital.

As part of PADETC's advocacy mandate, we, in collaboration with development partners, provide facilitation skills to the local communities that help them raise their voices and concerns in preserving and protecting community livelihoods. Out of this initiative, the story "Bye Bye Middle-Men, Hello Team Work and Profitability," about Mr. Songkao's family from the Khamu ethnic group in Beng district, Oudomxay, North Laos, was published (Cord.org 2015). Songkao's story promotes community voices and demonstrates how collective power through community partnership and teamwork can break the long-existing cycle of trade dependence on middlemen. In it, Mr. Songkao enthused:

> Together we are stronger and as a group, we are able to engage in new activities. Working as a group has also strengthened the relationships in the village; we trust each other and back each other when there are problems. The village as a whole has gained from this.

[3] https://www.youtube.com/watch?v=ivyqP0x-i1Y&list=UUmvPlWv9yj-PoOQyBt5eo-w&index=4 (Accessed April 8, 2019) This is a short video illustrating community-based restoration of a silted up local lake in Pukpeuk village, Xaybouli District, Savannaket Province. The intervention began with research and was followed by community action: weeding out debris in the lake, clearing the small canals, and establishing the vicinity as a conservation area. As a result of the action, the paddy fields are now receiving much needed water flowing through the canals.

In the nearly three decades since 1996 of being the first-ever Lao organization to focus on developing human resources for a sustainable development model, PADETC has gained extensive experience and reached important milestones. In 2005, PADETC was recognized as a local "champion" for empowering youth to be community leaders.[4] PADETC is keeping up this work with the development of civic groups and facilitating policy dialogue by engaging its beneficiaries in design and implementation at community levels.

At the moment, PADETC has executed four interconnected development programs in six provinces of Laos, namely policy advocacy, education for sustainable development, Lao CSO capacity development, and service delivery. Youth are positioned as central to these interconnected programs.

As part of this empowerment, we think that policymakers can do their part to remove legal and regulatory barriers that stand in the way of business innovation and investment. In April 2009, the Lao PDR created its first-ever legal frameworks for the existence of non-governmental organizations, namely Decree 115, or the "PM-Formulation of the Non-Profit Association Decree," and Decree 013, or the "PM-NGO Guideline." Both were endorsed in February 2015. They allow CSOs to operate and grant them an increased ability to participate in the political space. Above all, this is a clear recognition that responsibility for development does not rest with the state alone (Alagappa 2004, 25).

With time, civic groups can encourage the government to make more sustainable choices and facilitate community-based development models that combine the betterment of livelihoods for the poor with the protection of the biosphere, such as turning waste to energy (Campbell and Salli 2012). Again, we believe that successful community empowerment is essentially a collective engagement in which local CSOs and local government agencies as well as—ideally—the private sector are all equal.

PADETC continues to play an important role in empowering local community members by developing the training skills to develop strong community-based organizations (CBOs). These skills are in the areas of leadership, facilitation, project cycle management, and community action research.

Youth Empowerment

The initiative to engage youth in community livelihood improvement through training in life skills development, leadership and management skills, and field learning practices and exercises began in 2004, when the concept of discovery-based learning and the role of youth volunteers was officially accepted by the Ministry of Education. PADETC's work with youth has contributed not only in terms of community development and promotion of the environment but also in changing the lives of individuals and their understanding of development. This includes gaining confidence, being

---

[4] PADETC's founder received the Roman Magsaysay Award in 2005. See https://www.rmaward.asia/.

part of something bigger than themselves, and taking on leadership roles that prepare them for the future. Through diverse types of community services, youth can become effectively engaged in society. Accordingly, since 2009, youth have been PADETC's primary target group for training and have become a priority before reaching out to the broader community.

With assistance from development partners, PADETC designed a community development training course targeting young community members along with other training tools. By November 2018, a total of 331 rural youth with 109 females (aged 16–26) had successfully passed the 4-month intensive training course covering skills such as leadership, team-building, organic farming, village participatory research, gender analysis, environment management, communication, and development of relevant training of trainers.

After completing one-month field practicums in four villages in Xiengkhouang province and Vientiane capital, youth participants reflected their best learning outcomes as follows:

- Self-confidence is demonstrated during the process of interviews (asking questions) and the ability to make presentations of the findings in front of many adults.
- Increased capacity to think and develop a series of inter-related questions. It can be said that the basic foundations for critical thinking were established.
- With better knowledge of the living conditions and lifestyle of villagers, the youth became more concerned about the rural environment, increased their respect for the lifestyle and values of rural communities, became more aware of the difficulties faced by rural families, and became more respectful of local ethics and values.
- Youth expressed that they want to be more engaged in some type of development activities in their communities and in the communities that they have surveyed.

PADETC has documented the results of its youth capacity-building efforts. For instance, Ms. Thanouvong, a teacher in the Thongkhan Primary School, Vientiane capital, testified that "(now) I have seen progress in the young people. Before they were like children. (Now) they have learned how to manage their time, to be leaders, and followers, and to appreciate what it means to be a teacher; they have also developed value to help society."[5]

Today, about 35 former youth training participants are employed by Lao CSOs (Life Skills Development of Khong district, Champasack province; Mind Media; Rural Development Agency; Nasiew Vocational Training Center; and PADETC). Some are running small businesses within their communities.

[5] Personal communication, at the PADETC Fair on December 15, 2016.

Impacts on the Ground

Decree 115 of 2009 was issued by the Ministry of Home Affairs (MOHA) to create an enabling environment for the growth of Lao CSOs. It has gained traction, with approximately 160 NPAs and CBOs[6] currently registered, mainly for supporting the education and service sectors. This is an important window for civic groups to take on higher levels of responsibility and accountability vis-à-vis the implementation of the 8th National Growth and Poverty Eradication Support Program by 2020.

In recent years, the government has increasingly accepted that CSOs play an important role in supporting community development through service delivery and capacity building. Progress has been made relating to work with communities, for example in designing an appropriate package of technologies to support better farming practices (organic agriculture), marketing village products such as handicrafts, and identifying new possible enterprises.

Despite limited support from the government, today PADETC continues to support community empowerment through its small grant facility program, the provision of skills training, and coaching and mentoring in four target provinces (Savannakhet, Khammuane, Xiengkhuane, and Vientiane).

Ultimately, community groups have also begun to see themselves as capable of producing positive changes in their communities and are participating more. One of the major contributions of community engagement is that it encourages policymakers, in particular the local authorities, to see civic groups as valuable, untapped resources rather than as a problematic and marginalized portion of the population.

Between 2013 and 2014, PADETC embarked on multi-year, multi-stakeholder collaborations through projects such as the EU-funded Capacity Development for Citizen-led Inclusive Development (CD-CID), the EU-supported Forest Law Enforcement, Governance, and Trade (FLEGT), and Thaiban, a community-led research project supported by Oxfam Laos.[7] Through these close partnerships, PADETC has gradually developed trust with government counterparts, especially at the provincial level. This has resulted in the signing of four MoUs to open space for CSO-government collaborations on aspects of education and sports. The MoUs involved provincial education offices and governments, namely in Champasak, Savannakhet, Khammouane, and Xiengkhouang.

Factors that Shaped the Impacts

PADETC's presence has been seen in rural areas throughout the country for many decades and it is recognized as a local "champion" in terms of empowering community leadership. This work resulted in PADETC's founder (Mr. Sombath Somphone) being awarded the prestigious Ramon Magsaysay Award in 2005. Until today,

[6] NPA stands for Non-Profit Association. CBO stands for Community-Based Organization.

[7] Thaiban research was supported by Oxfam Laos during 2013–2014.

PADETC has continuously supported skills development for all civic groups (youth groups, community-based organizations, civil society organizations, and non-profit associations) to engage with community development. We have done this through the provision of regular training programs, the small grant facility, and direct service delivery (PADETC 2012).

Unfortunately, PADETC's founder Mr. Sombath Somphone was abducted on December 15, 2012,[8] and has since completely disappeared. Thereby until today, concerns about the government's perception and acceptance of civil society have remained. The government's position is becoming less clear, and despite the changing paradigm toward more inclusive governance, there remains the view that civil society "disrupts harmony."[9] As a consequence, the four key areas of PADETC's programs (Policy Advocacy, Education for Sustainable Development, Capacity building for Lao CSOs, and Service Delivery) have slowed down. In particular, the program on policy advocacy has almost completely stopped.

PADETC has a strong staff and team that stays together, is inspired, and has an effective communication and work style. In the course of a recent review of PADETC,[10] donors and partners confirmed that they maintain a strategic partnership with PADETC, therefore going well beyond the simple donor/recipient relationship. The review also confirmed that capacity development at the community level has remained PADETC's strength, as well as capacity development for CSOs, and with this, youth remain the driving factor throughout the process.

References

Alagappa, M. 2004. *Civil Society and Political Change in Asia: Expanding and Contracting Democratic Space*. Stanford, CA: Stanford University Press.

Campbell, B., and P. Salli. 2012. "Low-Carbon Yak Cheese: Transition to Biogas in a Himalayan Socio-Technical Niche." *Interface Focus*. https://doi.org/10.1098/rsfs.2012.0052.

Cord.org. 2015. "Bye Bye Middle-Men, Hello Team Work and Profitability." *Corg.ord* website. https://www.cord.org.uk/News/bye-bye-middle-men-hello-team-work-and-profitabiilty.

European Commission. 2009. "Commission Decision of 28 September 2009 Authorising Certain Member States to Revise their Annual BSE Monitoring Programmes." *Official Journal of the European Union*. L 256 : 35–37. EUR-Lex website. https://eur-lex.europa.eu/legal-content/EN/ALL/?uri=CELEX%3A32009D0719.

Gerin, Roseanne, and Ounkeo Souksavahn (translator). 2016. "New Lao Prime Minister Issues Ban on Timber Exports." *Radio Free Asia* website. https://www.rfa.org/english/news/laos/new-lao-prime-minister-issues-ban-on-timber-exports-05172016152448.html.

PADETC (Participatory Development Training Center). 2012. *Annual Report*. http://www.padetc.org/wp-content/uploads/2013/10/PADETC-2012-Annual-Report.pdf.

[8] See *The Enforced Disappearance of Sombath Somphone: A Documentary* at https://youtu.be/7-MmZUvMZWI

[9] Personal communication with a stakeholder. Also cited in PADETC Strategic Planning Document (2018–2025).

[10] PADETC Strategic Planning Document (2018–2025).

Schloenhardt, Andreas. 2017. *Trafficking in Persons from Cambodia, Lao PDR, and Myanmar to Thailand.* Bangkok: United Nations Office on Drugs and Crime (UNODC), Thailand Institute of Justice (TIJ). https://www.unodc.org/documents/southeastasiaandpacific/Publications/2017/Trafficking_in_persons_to_Thailand_report.pdf.

Sigsgaard, P. 2002. "Monitoring without Indicators: An Ongoing Testing of the MSC Approach." *Evaluation Journal of Australasia* 2 (1): 8–15.

Chapter 7
How Biodiversity and Culture Can Fuel Economic Prosperity: The Case of Traditional Textile Artisans of East Sumba, Indonesia

Chandra Kirana Prijosusilo

Highlights

- Indonesia's traditional textiles arose from indigenous people's profound knowledge of local biodiversity and how it could be nurtured sustainably to create woven cloth of extraordinary beauty and cultural significance.
- There is a huge potential market for traditional woven fabrics, as consumers increasingly take interest in supporting sustainably sourced clothing, yet this potential is largely untapped.
- An Indonesian NGO, Sekar Kawung, undertook to support East Sumba textile artisans through fair trade practices, helping them build confidence in their own knowledge and stories, and assisting them to access and work with new markets, all in spite of a number of challenges.
- Sekar Kawung engaged creatively to support the East Sumba artisans through documenting the complex production process, organizing exhibitions, a photovoice project for young people, and an ecotourism scheme; and integrating local artisanal knowledge into the formal education curriculum.

Introduction

In terms of biodiversity, Indonesia is one of the wealthiest nations in the world. This feeds into a rich cultural heritage fueled by the creative way indigenous communities use this biodiversity in their natural surroundings. From a biodiversity perspective, exploring Indonesia's traditional textile cultures is an amazing adventure. Different ethnic groups traditionally use various fibers and dyes and boast distinct patterns and motifs, often inspired by local biodiversity and sustainably utilizing natural materials.

C. K. Prijosusilo (✉)
Sekar Kawung Foundation, Sumba, Indonesia
e-mail: chandra.kirana@sekarkawung.com

M. Indrawan et al. (eds.), *Civic Engagement in Asia*,
https://doi.org/10.1007/978-981-16-9384-7_7

For millennia, people in the East Sumba Regency of the Indonesian island of Sumba have made bright red natural dyes for their textiles. These are derived from some twenty different plants that grow in the wild, including roots from the Noni tree (*Morinda citrifolia*) and candlenuts from the Kukui nut tree (*Aleurites moluccanus*). This use of plants reveals sound knowledge about the chemical principles involved in dyeing textiles. Modern research shows that species from the plant genus *Symplocos* hyper-accumulate aluminum (Al) in their leaves, making them an important mordant. Mordants are crucial to the dyeing process, as they ensure that textiles are colorfast. Amazingly, centuries before the development of modern analytical chemistry, people in East Sumba were already selecting and using *Symplocos* sp as a mordant (Cunningham et al. 2011). Candlenuts contain high levels of oil, which also makes them a useful mordant. The oil is applied by soaking threads in a nut paste before soaking them in the red dye preparation (Ingram, n.d.).

Today, a growing number of consumers are consciously seeking sustainable products. Here, the indigenous approach to production could potentially bring high economic returns and create sustainable economic prosperity. However, this potential is mostly untapped and indigenous knowledge is being eroded along with the wealth of biodiversity in their lands. Without any intervention, within less than a decade from now (2019), many indigenous approaches to sustainable production could become extinct, causing the knowledge about biodiversity and various sustainable production methodologies to disappear along with them.

In 2013, our NGO, Sekar Kawung, began working with Paluanda Lama Hamu, a group of textile artisans who make *tenun ikat* (woven cloth), to open opportunities in sustainable markets by fairly trading with them.[1] This chapter will discuss this work as an illustrative case of how diversity and culture can fuel economic development. After providing more contextual information on the Paluanda Lamu woven cloth workers of East Sumba in the next section, the chapter will explain how confidence in the Paluanda Lamu Hamu people's story has been built through several project activities led by the NGO Sekar Kawung. It will then discuss attempts to reach out to new markets, for example by using the Internet and a textile art exhibition. The chapter will then move on to analyzing these experiences, drawing out some lessons learned, as we take stock and look to what the future holds for Paluanda Lamu woven cloth.

[1] This group of traditional textile artisans was set up in the early 1990s by a woman named Agustina Kahi Atanau (familiarly known as Mama Dan) to enhance economic opportunities for the traditional textile artisans in her clan. The meaning of Paluanda Lama Hamu is "walking together in a good direction."

The Paluanda Lama Hamu Woven Cloth Artisans of East Sumba

Most Paluanda Lama Hamu members live in the villages of Lambanapu and Mauliru, in Kambera subdistricts, in East Sumba. Lambanapu and Mauliru people still wear traditional hand-woven textiles for community gatherings such as the birth, marriage, or death of a community member. During these events, people gift textiles to each other following traditional rules, which have been handed down over centuries. Therefore, not all textiles are produced for sale, and a significant amount of woven cloth remains within communities.[2] This is the cultural setting where the Paluanda Lama Hamu group members live: they fulfill their cultural and livelihood obligations by producing more woven cloth than they need for themselves, selling the surplus to outside consumers. Many Paluanda Lama Hamu members do not own much land and fully depend on making woven cloth for their livelihoods.

During 1980–2000, the uniquely pictorial and extremely complex Eastern Sumbanese woven cloth experienced a booming trade dominated by traditional oriental folk art collectors (Forshee 2001).[3] This market flourished thanks to the international tourist trade on Bali (Sekar Kawung 2017a). Paluanda Lama Hamu learned to navigate this market and thrived. But in 2002, after the Bali bomb incident that killed 202 people in the tourist district of Kuta, the boom came to a standstill. As a result, during 2002–2013, Paluanda Lama Hamu struggled to survive.

As times have changed, Paluanda Lama Hamu is now being challenged to create market-savvy products to sustain their livelihoods. Often this means adapting to fulfill contemporary needs that may be alien to them. As an NGO, we at Sekar Kawung needed to figure out how to help them do this, while at the same time ensuring that their cultural distinctiveness was maintained.

Enabled by a substantial grant from Millennium Challenge Account Indonesia (MCAI), through the intermediary agency of the Samdhana Institute, Sekar Kawung engaged with Paluanda Lama Hamu during June 2016-March 2018 to jointly design and implement a program that aimed to:

- Strengthen their cultural identity as *tenun ikat* (woven cloth) artisans;
- Strengthen their market positioning in the ethnic arts market and make a strong entrance into the sustainable consumption market;
- Initiate and nurture new economic activities to provide non-weavers with sustainable livelihoods, which would, in turn, support a woven cloth-based economy to thrive;
- Promote the restoration of degraded lands using natural dye trees and plants.

Aspects of this work are explored throughout the rest of this chapter.

---

[2] For a deeper insight into the meaning of traditional textiles in East Sumba, see Adams (1969).

[3] Full copy readable at https://epdf.tips/between-the-folds-stories-of-cloth-lives-and-travels-from-sumba.html.

The Diversity of the Woven Cloth: Symbols

Each piece of woven cloth is essentially a work of art. No piece is the same—each is the unique fruit of its designer's creativity, imagination, and skill. Textile motifs and designs in traditional cultures almost always carry deep philosophical meaning, but what exactly are these meanings? To answer this question, Sekar Kawung engaged the artisans in creating a woven cloth symbol dictionary. We worked together to decipher every single symbol we found on 50 pieces of Paluanda Lama Hamu woven cloth. When an artisan could not explain the meaning of a symbol, we would ask an older artisan. The exercise resulted in the documentation of 41 symbols into a symbol dictionary. This dictionary was used to read 50 pieces of Paluanda Lama Hamu traditional woven cloth tales. From these, we learned there are three types of woven cloth based on their content, which are:

- A message of blessing or prayer for well-being, dignity, and success in life;
- Advice or teachings on how to be a good human being;
- Recordings of stories, especially stories of royal burials.

This exercise created confidence within each artisan, helping them consciously know and feel that they were true artists. It became clear that each woven cloth designer had a distinctive visual style, even if they were using the same symbols on their pieces of textile. Each designer began to speak about their creations more confidently and enthusiastically, explaining where their inspirations came from and what messages they wanted to relay through their work.

Some excerpts from the symbol dictionary are illustrated and explained below (Figs. 7.1, 7.2, 7.3, 7.4, 7.5, 7.6, 7.7, 7.8, 7.9 and 7.10).

Fig. 7.1 *Andung* (skull pole), also named *pohon hayat* or tree of life, is a symbol of victory and greatness. These chandelier shaped poles are adorned with the heads of the enemy

Fig. 7.2 *Atanau* is a human body with animal body parts. His message is that humans must protect all beings in nature and live by enriching the other. Some artisans say that in *Marapu* belief, the *Atanau* is the symbol of sincerity and faith. When the *Atanau* is pictured with its hands open, it symbolizes having nothing to hide in the presence of God

Building Confidence in Their Own Story

Given the rich cultural heritage of Paluanda Lama Hamu, Sekar Kawung was committed to helping group members build confidence in their own stories. Therefore, Sekar Kawung combined learning about the market with an exploration of the East Sumba textile tradition in Lambanapu and Mauliru (Waingapu Subdistrict) from within. The community was taken on an adventure to deeply review their own culture through various activities and approaches, such as documenting the creative process, creating a photovoice for children, establishing an ecotourism scheme, and integrating woven cloth into the formal education curriculum. Aspects of these are described below.

Documenting the Creative Process in the Tenun Ikat *Value Chain*

It takes 42 steps to make a *tenun ikat* piece and up to twelve people to complete it. The intricate creative process demands high-level skills, which the Paluanda Lama Hamu artisans know by heart. If these artisans want buyers to understand where they are coming from, they must be able to tell the story behind the products to customers. So this was a journey of recounting and documenting, as well as a journey of trial and error, to explore innovations and experiments to create new products through the craft of traditional woven cloth (Fig. 7.11).

The exercise helped us to understand existing social economy structures in Lambanapu and Mauliru. We were able to identify specific trades that were alive

Fig. 7.3 *Wuyarara*—The crocodile is the symbol of royalty for a king and kingly courage

in the community, and how they all interconnected in the making of woven cloth. This, then, enabled us to develop clearly targeted forms of assistance and appropriately design new potential economic activities for the community. The process showed how a strong woven cloth culture could function as a good foundation that would allow different village-based economic activities to develop independently, and yet still be connected and strengthen each other.

The table below explains the different trades in the woven cloth value chain, the contributions and roles of each trade, and the new economic possibilities offered by each.

Fig. 7.4 ***Habak***—This symbolizes a hard-working woman and beauty. Sometimes it is encircled with a wreath to symbolize kinship and human interrelationship

Fig. 7.5 ***Kambiha ndjara***—The hoofprint of a horse symbolizes prosperity

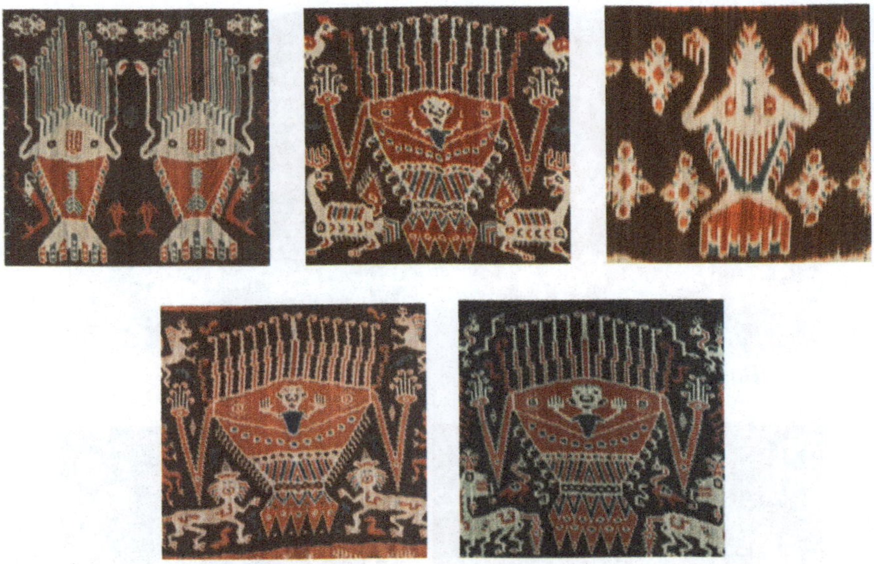

Fig. 7.6 ***Kandala***—This symbolizes life after death and incarnation

Different trades in the *Tenun Ikat* value chain	Contributions of each trade and the roles they play	New economic possibilities
Yarn Makers	Grow semi-wild cotton in their gardens to make yarn. Their families for generations traded yarn in the village	There is a market for naturally colored hand-spun cotton. Yarn makers can also become teachers for tourists, participating in cultural tours
Tenun Ikat Designers	Sell drawing skills; help people create the designs they want on *tenun ikat* pieces. The owner (a local person or local village investor who owns the yarn and would like designers to design the cloth) explains what symbols they want, and the designer composes the combination of symbols on the yarn, which has been prepared within a frame to resemble a canvas that can be drawn upon. Historically designers directly tied the yarn in constructing their design by heart as they went along. Today, most designers use colored pencils, and the tying process is carried out by skilled helpers. Designers are respected and highly paid in the business of *tenun ikat* production	Need to be strengthened to become owners of their pieces instead of only working for investors. This will enable them to build a reputation as *tenun ikat* artists with end consumers

(continued)

(continued)

Different trades in the T*enun Ikat* value chain	Contributions of each trade and the roles they play	New economic possibilities
Blue Color Experts	The blue color expert works alone. They are considered to have special powers handed down from ancestors. Traditionally, only specific families are recognized as blue color experts in Lambanapu and Mauliru. Many taboos are involved in carrying out their work. They are indispensable. The whole village relies on them. They are highly respected and well paid. Traditionally, only women are permitted to work with this color, but this is gradually changing	Can expand their trade to service other types of textile production, for example, *shibori* makers. They can also sell surplus indigo paste at a good price
Red Color Experts	Men and women do red coloring in Lambanapu and Mauliru. This trade is well paid	Can expand their trade to service other types of textile production, such as *shibori* cloth makers. They can also sell surplus powdered mironda root
Tenun Ikat Weavers	Only women weave. *Tenun ikat* is only done by highly skilled weavers. They must ensure the designs on the threads are put together neatly and evenly in the weaving process	Weavers can make all sorts of products other than putting together the *tenun ikat*
Kabakil Weavers	Kabakil is a patterned weave applied to both ends of a high quality *tenun ikat* cloth known as the *hinggi*. Not all weavers can weave this part of the *tenun ikat*. Therefore, they provide a special service to the entire *tenun ikat* industry within a village	The *kabakil* can be made into jewelry, and small products such as camera straps, hair bands, and so on. *Kabakil* artisans do not need to depend on the *tenun ikat* alone for their livelihood
Supporters	Provide simple yet important skills in assistance throughout the entire *tenun ikat* making process. Skills such as rolling the yarn into balls, applying the yarn to frames before the design process, or practicing the art of tying the yarn to resist the dyes in the coloring process	As they have very good insight into the whole *tenun ikat* making process and are well connected to all the different experts in the field, they have the best knowledge needed to run cultural tours for visitors

Thousands of beautiful photos that deeply reflect the makings of woven cloth were created in this exercise. Outreach material to promote woven cloth was disseminated through exhibitions, Instagram, and Facebook. One outstanding piece of promotion born through this process is the short film *Tenun Ikat Sumba Timur* (The Woven Cloth of East Sumba) by Sekar Kawung (2017a). It beautifully tells the story of woven cloth-making, going beyond the weaving to illustrate the deep relationship

Fig. 7.7 *Naga*—The dragon is a symbol of bravery and victory

the people have with the landscape and plants surrounding them. It shows how East Sumbanese people make their clothes by using only very local materials.

A Photovoice Project for Children and an Ecotourism Scheme

Twenty children aged 10–14 years old were selected from two local primary schools to take part in a photo project. They were each given a camera and were taught how to use it, as well as encouraged to think of how photography can be used to

Fig. 7.8 *Njara*—The horse is a symbol of authority and grand prestige

communicate what they felt was important to the world. This was combined with trips around the village to hunt for photograph opportunities. Thousands of photos resulting from these trips were presented by the children and discussed together. The children chose two or three of their favorite photos and learned how to write captions for these selected pictures for an exhibition.

This process was led by an experienced photographer in collaboration with Sekar Kawung's three field staff. Fifty photographs were curated and presented to professional gallery standards. An outdoor exhibition was held on the grounds of the local Catholic primary school in Lambanapu for three days, accompanied by nightly cultural presentations of dance, poetry, and traditional music. All the children photographers hosted the exhibition, taking hundreds of visitors through the exhibition to

Fig. 7.9 ***Rusa***—The deer is a symbol of high status. Some use it as a symbol of nature conservation

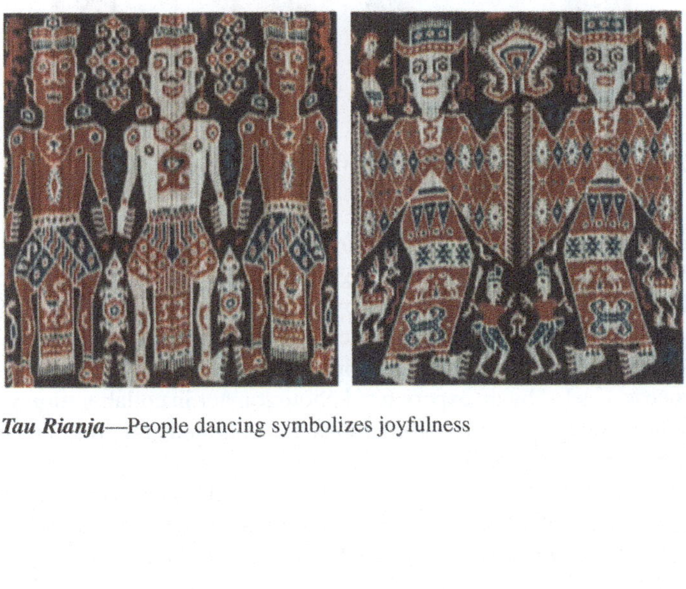

Fig. 7.10 ***Tau Rianja***—People dancing symbolizes joyfulness

Fig. 7.11 New products were designed to suit the needs of consumers looking for wearable and usable items with a story that spoke to environmental consciousness and cultural respect

enjoy the pictures. In the opening ceremony, the children were presented with a book consisting of their pictures and the stories related to each picture. The book was titled: *Cerita Lensa Hati Kami Anak Kandung Tenun Sumba,* or *Stories Through the Lenses of Our Heart: The Children of Sumba Woven Cloth* (Sekar Kawung 2018).[4]

We gained some important insights from the exhibition, which helped to plan some follow-up action. This included an appreciation of the fact that making the woven cloth is not a purely female activity, and so any effort to improve the economy and sustainability of woven cloth-making needs to take on board both male and female perspectives and ensure equitable benefits for both genders. Another insight was that the children's photographs revealed the richness of wildlife diversity in the two villages, and this led to the development of an ecotourism scheme to offer to visitors (Fig. 7.12).

Overall, the exercise enhanced the appreciation of traditional woven cloth and the local endowment of biodiversity among the younger generation in the project area (Sekar Kawung 2016a, b), and confidence in the village's potential as an eco-culturally responsible tourist destination.

Integrating Tenun Ikat *into the Formal Education Curriculum*

In collaboration with the local bureau of education, we facilitated a multi-stakeholder process involving representatives from the local government planning agency and parliament, primary school principals, primary school inspectors, teachers, students, and senior artisans as the main resource people to develop a special module about *tenun ikat* for fourth graders. Multiple workshops prepared draft modules and two

[4] See also a video on this project entitled, *Menenun Harapan di Tanah Sumba* [Weaving hope in the land of Sumba], from minute 3:52 to minute 4:37 (Sekar Kawung 2018).

Fig. 7.12 *Petualangan Anak Kandung Sumba* meaning Adventures of Sumba's children is the second book the children wrote, it is about the wildlife in their landscape. Some of the authors at the photo exhibition that was held to launch the first book entitled *Cerita Lensa Hati Kami Anak Kandung Tenun Sumba*, or Stories Through the Lenses of Our Heart: The Children of Sumba Woven Cloth.

trials were carried out, followed by policy discussions to ensure the module could be fully integrated into the formal education system and the district's education budget.

The module consists of:

- *Teacher Book: East Sumba Tenun Ikat*
- *Student Book 1: The Cultural History of Sumba Textiles*
- *Student Book 2: The Process of Making Woven Textiles*
- *Student Book 3: The Natural Dye Process*

The Director-General of Culture and Education of Indonesia endorsed this module, and the local parliament with the local development planning agency in East Sumba have given their formal support. In 2018, the local government supported a pilot program to teach the module in 20 selected schools, intending to upscale this into all primary schools in East Sumba by the 2019–2020 school year.[5]

Reaching Out to New Markets: Moving Beyond Comfort Zones

Using the Internet to Understand the Market

To gain inroads into new markets, Sekar Kawung selected nationally and internationally conscious consumers committed to sustainability as the target market for East Sumba textiles, because our values of protecting nature and enriching culture

---

[5] All four books are available online: Student Book 1 at http://bit.ly/2BTC5hj Student Book 2 at http://bit.ly/2CiI5vI Student Book 3 at http://bit.ly/2HcHyPq and the Teacher Book at http://bit.ly/2BVqHl4.

while building economic prosperity are aligned. To help the artisans in Lambanapu and Mauliru understand this market we used the Internet.

We took Paluanda Lama Hamu artisans on adventures across Instagram, Pinterest, and online shops like Shopify and Etsy to see how consumers all over the world were seeking and responding to clothes, as well as home and living products made from traditional textiles, including many from East Sumba woven cloth.

Using beautiful photos from these Internet sites, Sekar Kawung staff explained what conscious consumers were looking for as well as various important concepts, such as slow-fashion, hand-made textiles, natural dyes, and most importantly, the growing desire among consumers to know the faces of the people who made the clothes they were purchasing. The Internet showed that consumers were interested in knowing how things are made, and how this impacts the environment and the people involved in making them.[6]

A Google search for traditional textile "wearable art" delivered a rich repository of traditional textiles images, many from East Sumba, enabling discussions about identity in terms of being important contributors to the global market. Seeing that traditional textile pieces were seen as wearable art on the Internet helped them to rethink their self-concept and self-esteem. Seeing *tenun ikat* on international fashion runways and decorating luxurious lounge rooms and other interior settings instilled a sense of pride born by the knowledge that the products they were capable of making were highly valued by consumers from all over the world.

Exploring the Internet also helped them to understand new market challenges. This illuminated the reality that the market for traditional textiles was moving away from the world of collectors and the global "ethnic arts" markets and into the fashion and home and living markets. It became clear that the larger pool of consumers were millennials looking for wearable and usable items with a story that spoke to environmental consciousness and the human dignity of the makers (Gonzalez 2015). This is a story that Paluanda Lamu could tell, because it is a story they are living in their daily lives.

Some Example Initiatives

To reach out to new markets, we proceeded with three initiatives. First, we held a textile art exhibition in the Bank Mandiri Museum of Jakarta. We chose a bank

[6] See, for example, the Red Carpet Green Dress, a sustainable fashion campaign founded by Suzy Amis Cameron as part of fundraising for MUSE School CA at https://www.instagram.com/redcarpetgreendress/ and Emma Watson's Stylish Guide to Shopping Sustainably, at https://www.whowhatwear.com/emma-watson-ethical-fashion.

museum to highlight the important effect of banking on culture for sustainable development. Exhibition visitors were able to explore the *tenun ikat* creative process through a combination of installations, movies, curated art pieces, posters, and interaction with textile artisans (Sekar Kawung 2017b). Seven Paluanda Lama Hamu designers presented their art, each with a dedicated gallery portraying their works. They were accompanied by seventeen other artisans who also participated in the exhibition. We recorded 1,576 visitors in 8 days, with 677,364,000 IDR (over 48,000 USD) in sales by Paluanda Lama Hamu.

Second, we further engaged in social media and offline collaborations. From its inception, the project actively used Facebook, Instagram, and YouTube to promote information and knowledge about woven cloth in Lambanapu and Mauliru. This helped to connect with many important links that endorsed Lambanapu and Mauliru villages as quality *tenun ikat* centers in East Sumba, and as interesting tourist destinations (e.g., Sekar Kawung 2016a, b). Social media helped inspire millennials about East Sumba woven cloth. Many millennials visited Lambanapu and Mauliru while the project was still ongoing, connecting directly with woven cloth artisans. Through social media, Sekar Kawung became acquainted with fashion and design students from universities in Indonesia and Singapore. Many began using us as a resource on East Sumba *tenun ikat* issues. Young university and design students offered invaluable voluntary help in preparing and running the art exhibition.

Third, we built capacity and new economic initiatives. Using the woven cloth as the central economic driver, the project built the following: a Tenun Ikat Village Gallery; a Tenun Ikat coloring workshop; a three-bedroom guesthouse; a small Tenun Ikat cultural tour business; a natural dye plant nursery and biodiversity education center; and a local catering initiative.

Challenges and Solutions

The approach used in this woven cloth endeavor draws from the principles used in traditional agroforestry systems throughout Indonesia. In this approach, communities internalize and practice the knowledge of the different elements of local biodiversity (Michon and de Foresta 1996). Traditional farmers in Indonesia have used this knowledge to design their natural farming production systems with minimum maintenance and high yields for centuries. They can do this because they take what is strong in nature, and nurture it in its context, instead of decimating everything in a landscape to plant one variety of a plant that is considered valuable. In modern agriculture, a similar approach can be found in the permaculture movement.[7] The author of this chapter grew up on a traditional Javanese farm, where the *pekarangan*, or homestead area, was a lush agro-forest system and witnessed this productivity daily. Her grandmother was a medicine woman, and she would make medicine and cosmetics from ingredients harvested daily from this area. All food materials, excluding rice,

[7] What is Permaculture, www.permaculture.com.

came from this area, as well as building materials, and some natural dyes for her batik making. There was always much surplus to sell. This differed very much from the monoculture approach in modern agriculture that is now dominating markets and landscapes. Traditional farmers had a large variety of products to sell at different times of the year. They had daily, weekly, seasonal, and yearly produce to sell. Modern farmers have a large amount of one variety to sell, all at the same time.

The biggest problem faced by traditional farmers today, including the woven cloth artisans in East Sumba, is a serious detachment from markets. What this intervention tried to do was to strengthen the rich culture of the woven cloth artisans in East Sumba and connect it to the growing communities of niche markets in the world that value diversity and aim for sustainability and justice. These new markets contain an ever-expanding plethora of sustainability-conscious small market niches. There is a parallel in this market to the way traditional communities manage their lands and their ways of production.

Other challenges we faced during the project related to funding approaches and donor relationships. We came to realize that a "one size fits all" approach to funding could work against our aims. For example, synchronizing activities with the time that the donor transferred the funds was not always possible, so sometimes funds would be returned unspent. There were difficulties with attaining land certification and local government permits for building, especially those needed to ensure that donor and central government environmental and social standards were fulfilled. For example, land to build upon needed individual ownership certificates, yet land in the project area was owned communally. We were not convinced that certain official performance standards that needed to be implemented were genuinely for the sake of the environment and the community, rather than for administrative purposes only. In terms of personnel, the stresses of the project took their toll as well. Negotiating between donors, builders, the local communities, and other similar organizations was a balancing act, and we did not always succeed. There are lessons to be learned from all of these situations.

Taking Stock and Looking to the Future

Taking stock of life for the Paluanda Lama Hamu artisans and the community of Lambanapu and Mauliru, and the central players in the project from Sekar Kawung at time of writing (January 2019) there are several areas we wish to highlight.

Tenun Ikat Paluanda Lama Hamu

At the end of the project, Paluanda Lama Hamu was split into two groups (because they turned out to have different working rhythms). Kornelis Ndapakamang set up

Tenun Ikat Prayikundu as an offspring of the original organization (Paluanda Lama Hamu).

Paluanda Lama Hamu is now based in the Tenun Ikat gallery, which has been named Uma Hondung and is now led by Theresia Mbati Mbana. It has hosted many education sessions for Mai La Huma tours, teaching visitors all the different steps in the *tenun ikat* process. It also provides blue coloring services to new consumers and sells a surplus of dried indigo paste. Its members also use the gallery as their shop, where all the woven cloth is sold.

Sekar Kawung is back to a trading relationship with Paluanda Lama Hamu, ordering pieces of textiles to make into various fashion items. New buyers have also been approached, such as Etnickhatulistiwa, a creative slow-fashion social enterprise based in Bogor, and the Gerai Nusantara shop, which specializes as a department store for indigenous peoples' products. Sekar Kawung still facilitates international sales for Paluanda Lama Hamu, and pieces have been bought by buyers from Australia, the USA, and Japan. This is still a very slow process, with the potential to grow in the future.

Tenun Ikat *Prayikundu*

Kornelis Ndapakamang, previously the head of Paluanda Lama Hamu, is a passionate *tenun ikat* designer and craftsman. After the *tenun ikat* textile art exhibition in Jakarta, he has responded positively to many opportunities coming from new partners. So far, he has participated in international textile exhibitions in Malaysia, in a World Ikat Textile Heritage Exhibition, as well as in a high-end textile exhibition in a high-end resort in Bali in December 2018.

Our experience with Kornelis shows that in the case of East Sumba *tenun ikat*, helping one textile art designer to thrive in the international textile art scene can have a big impact in creating sustainable jobs and sustainable land management in the village. This is because one designer needs all the artisans who are masterful of all 42 steps to help him or her make a piece of *tenun ikat.* Designers are also the most powerful point of influence in tree planting. Kornelis avidly campaigns for the planting of natural dyes and will buy resources from neighbors far and near who are planting these crops instead of harvesting them from the wild. Future plans with Kornelis will be to promote planting more cotton in Lambanapu and Mauliru (Fig. 7.13).

When Sekar Kawung set up an office for the project in Waingapu, we connected Jakarta-based media with Paluanda Lama Hamu. Since then, by combining masterful artwork with a sharp acumen for media, he has been able to very positively enhance his brand, Tenun Ikat Prayikundu, and East Sumba *tenun ikat* in general. Most recently,

Fig. 7.13 Master Weaver Kornelis Ndapakamang, Sumba Woven Textiles Exhibition, Bali, 2018

he was featured in a CNN Indonesia TV program named *Inside Indonesia* featuring *Menganyam Cerita Tenun Ikat Sumba* (Weaving the Story of the Sumba cloth).[8]

Since the project closure, Kornelis and his wife Desmiyati Hambabanyu have sharply increased their flow of international visitors and have done so independently. They often refer these visitors to Mai La Humba.

Mai La Humba Tours

Mai La Humba tours began with four people in the last quarter of the project. Although two staff have found more secure employment elsewhere, the tours continue to steadily grow stronger. The bike village tours to learn about *tenun ikat* culture are loved by those who partake in them, and they are well integrated into the *tenun ikat* artisan network in their village. Sekar Kawung recently received feedback from a group of thirteen Australian visitors who said they had a great experience. Monitoring of Facebook and Instagram reveals that there is a small but constant flow of visitors coming to learn about *tenun ikat* in Lambanapu and Mauliru on a Mai La Humba tour. They often receive visitors referred to them by satisfied prior visitors and are linked with Wise Step Travel, a responsible travel tour operator based in Jakarta.[9]

Jaladwara, the responsible tour practitioners and consultants that Sekar Kawung hired to train youth in eco-culture tourism during the project, on their initiative, continues to provide technical assistance to Mai La Humba within their financial

[8] https://www.facebook.com/kornelis.ndapakamang.7/videos/1021804967988286/UzpfSTEwMDAwNDc2Mzc3MjM1ODoxMTEwODY2NTAyNDE1NDY1/.

[9] https://www.wisestepstravel.com/.

capacity, sometimes raising funds to do this. Currently, Mai La Humba is asking for assistance to improve their business management capacity.

Sekar Kawung

Sekar Kawung is back to being a trading partner with Paluanda Lama Hamu and Prayikundu. We specifically target the poorest members, providing them with quality yarn and working with them to create designs for our hand-made eco-fashion line. We sell online and in special bazaars. Our online shop can be found at www.sekarkawung.com.

Additionally, we now trade with *tenun ikat* artisans outside of Lambanapu and Mauliru, especially the poorest of the poor migrant weavers who originate from Sabu and have unique Sabu *tenun ikat* motifs but do not use natural dyes. We connect these weavers with Sumbanese natural dye experts, enabling the production of beautiful naturally dyed Sabu *tenun ikat.*

Recently, Sulwhasoo, a leading Korean cosmetic company that highly values biodiversity and culture conservation, approached Sekar Kawung to collaborate with them in a Corporate Social Responsibility project. The theme of the project, valued at 100 million IDR, is to nurture "Beauty from Your Culture."

Sekar Kawung is now working to create sustainable yarn for Indonesian textile artisans. Our latest project involves sourcing wild silks and encouraging the planting of semi-wild cotton that some indigenous communities are still planting. We are also seeking to develop a system that will be able to deliver these sustainable yarns as well as sustainable imported yarns to traditional textile artisans in Indonesia. We aim to put East Sumba's artisanal textiles squarely on the international sustainable fashion scene.

Conclusion

This chapter discussed the case of Paluanda Lamu *tenun ikat* (woven cloth) workers in East Sumba, Indonesia, as a way of considering how biodiversity and culture can fuel economic prosperity. It provided contextual information on the Paluanda Lamu woven cloth workers of East Sumba and showed some examples of their work. It considered how confidence in the Paluanda Lamu people's story has been built through several projects led by the NGO Sekar Kawung, and discussed attempts to reach out to new markets, for example by using the Internet. It also explored these experiences, the challenges that arose, and the solutions found to these challenges. Finally, it analyzed these experiences, drawing out some lessons learned, pointing to the future.

To bring the reader right up to date, since the end of the project (January 2019), several new developments and continuations of the work have taken place. Firstly, we were able to strengthen the cultural identity and self-esteem of Paluanda Lama Hamu *tenun ikat* artisans, thus fueling their creative passion to make masterpiece crafts. This helped to ensure the quality that sustainable fashion and living niche markets are looking for and seems to have reignited interest among collectors. One avid collector, Lewa Pardomuan, the chair of Wastra Indonesia, says that the Paluanda Lama Hamu Textile Art Exhibition in Jakarta made him realize even more than before that to keep the *tenun ikat* culture alive, it is important for collectors to purchase and value new pieces of cloth as much as old ones.[10] Through the textile art exhibition, combined with the Sekar Kawung online shops, we have been able to feature and strengthen the Paluanda Lama Hamu artisan group in the ethnic arts and sustainable consumption market as trusted artisans who deliver high quality naturally dyed East Sumba textiles.

The project has also helped the community to build a culture-tourism destination in their village, as we helped to establish a set of complementary economic initiatives. These included the homestay, the natural dye biodiversity educational center, the *oleh-oleh* home industry, and Mai La Humba tours. Some are growing; some others still need attention in terms of community development facilitation services.

Environmentally, the Biodiversity Education Center and Plant Nursery created by the project has failed and many of the thousands of seedlings distributed broadly to 87 local farmers outside of the Paluanda Lama Hamu group have died. One year after the project has closed, however, we can observe that the market for naturally dyed *tenun ikat* from Lambanapu and Mauliru is growing. With this, more people are independently planting, harvesting, and selling natural dye resources to important players like Kornelis Ndapakamang. In his words, "I do not preach, but I buy natural dye ingredients from people provided they are harvested without killing the trees, they then begin to plant, it gives them an income, they are noticing that it pays to plant natural dye trees." The lesson learned here seems to point to the notion that in cash-strapped communities, it is important to be able to apply almost immediate financial benefits to land rehabilitation efforts.

Finally, while this project has resulted in positive changes on the ground, the price paid by Sekar Kawung as the community facilitator was very high. Its heavy burden of administrative requirements and rigid funding mechanisms, especially in terms of scheduling transfers and expenditures, derailed Sekar Kawung from any meaningful activity in 2018, that is, after the project was closed. We lost our little café that sold community sustainable food products, and our small textile and fashion gallery. We also lost most of our staff as well as precious time to plan the year ahead and fundraise. Fortunately, we received a recurring small grant from UNICAL.[11]

In our view, if SDG targets are to be achieved, then funding mechanisms must be designed to consider the diversity of nature and culture, especially in the case of land-dependent communities. We think this is possible, provided that creative thinking is applied concerning systems in nature and systems in cultures. Financial accounting and administration mechanisms need to somehow adopt a systems thinking approach.

[10] http://wastraindonesia.org/.

[11] https://unicalcares.org/.

One year later, our online shops selling sustainability-conscious fashion are back on track. Not only this, but our confidence is also stronger. The foundation for sustainable economic development in Lambanapu and Mauliru is stronger because Sekar Kawung staff were mindful of how the local ecosystem and local culture influenced peoples' decisions. The team worked with their hearts, staying focused on the local rhythms of change even through long periods when funds from donors were delayed. This has resulted in the rekindling of top-quality intellectual knowledge among the community. Together, we revitalized the cultural spirit and self-esteem of the village. We are confident this will last and help the economy continue to grow in the future.

References

Adams, MJ. 1969. "Systems and Meanings in East Sumba Textile Design: A Study in Traditional Indonesian Art." Cultural Report Series, No. 16. New Haven: Yale University Southeast Asia Studies.

Cunningham, AB., Maduarta, IM., Howe, J., Ingram, W., and Jansen, S. 2011. "Hanging By a Thread: Natural Metallic Mordant Processes in Traditional Indonesian Textiles." *Economic Botany* 65 no. 3: 241–259.

Forshee, J. 2001. *Between the Folds: Stories of Cloth, Lives and Travel from Sumba.* Honolulu: University of Hawaii Press.

Gonzalez, N. 2015. "A Brief History of Sustainable Fashion." Triple Pundit website. https://www.triplepundit.com/story/2015/brief-history-sustainable-fashion/58046.

Ingram, William. n.d. "In a Lather over Candlenut Oil." Threads of Life. Accessed February 8, 2020. https://www.threadsoflife.com/in-a-lather-over-candlenut-oil/.

Michon, G., and de Foresta, H. 1996. "Agroforests: An Original Agro-Forestry Model from Smallholder Farmers for Environmental Conservation and Sustainable Development." In Proceedings of the UNESCO—University of Tsukuba International Seminar on Traditional Technology for Environmental Conservation and Sustainable Development in the Asia-Pacific Region, edited by K. Ishizuka, S. Hisajima, and D.R.J. Macer (held in Tsukuba Science City, Japan, 11–14 December 1995): 52–58.

Sekar Kawung. 2016a. "Cerita Lensa Hati Kami Anak Kandung Tenun Sumba" (Stories Through the Lenses of Our Heart: The Children of Sumba Woven Cloth). Photovoice. http://bit.ly/2nVUktK.

Sekar, Kawung. 2016b. "Petualangan Anak Kandung Sumba" (Adventures of the Children of Sumba). Photovoice. http://bit.ly/2o3TXML.

Sekar Kawung. 2017a. Tenun Ikat Sumba Timur (Woven cloth of East Sumba). Video. https://www.youtube.com/watch?v=oTxTcKuyhyI.

Sekar Kawung. 2017b. Pameran Tenun Ikat Sumba Timur di Museum Bank Mandiri Jakarta (Exhibition of Sumba Woven Cloth at Bank Mandiri Museum Jakarta). http://bit.ly/2EFL7iY.

Sekar Kawung. 2018. Menenun Harapan di Tanah Sumba (Weaving Hope in the Land of Sumba). Video, minute 3:52 to minute 4:37). https://www.youtube.com/watch?v=pHdPhFNC-No&t=500s.

Chapter 8
Localizing the Sustainable Development Goals: The Case of an Urban Community in Quezon City, Philippines

Nestor T. Castro

Highlights

- Research and education about SDGs can be implemented at the local level (localized) within particular communities.
- Community members can prioritize which SDGs are more relevant to them when local contexts are recognized. As part of culturally sensitive governance, it is therefore important to tap into local cultural values.
- Universities can play a role as facilitators to localize SDGs.

Introduction

In the past, the Philippines' strategies for the implementation of the Sustainable Development Goals (SDGs) were only developed at the national (see Appendix in this chapter), and in some cases, sub-national levels. There were no clear-cut strategies on how to implement measures at the local level to be able to achieve the SDGs, such as by local government units and by community members themselves. It is therefore important to identify local conditions and engage communities. This case study looks at an attempt by an urban community in the Philippines, i.e., Barangay UP Campus, to localize the SDGs. This was part of a research project of the Promotion of Sustainability in Postgraduate Education and Research Network (ProSPER.Net) entitled "Development of a Framework for the Local Implementation of the SDGs," undertaken in 2018–2019 in selected communities in India, Thailand, the Philippines, and Japan.

N. T. Castro (✉)
University of the Philippines Diliman, Quezon City, Philippines
e-mail: ntcastro1@up.edu.ph

M. Indrawan et al. (eds.), *Civic Engagement in Asia*,
https://doi.org/10.1007/978-981-16-9384-7_8

The Setting: Barangay UP Campus

The attempt to localize the SDGs relies heavily on the particularities of each local context. This section briefly outlines the nature and the complexities of the local community that provided the context for this research. Barangay UP Campus is one of the 142 *barangays* (villages) of Quezon City, a highly urbanized city within Metro Manila, the National Capital Region of the Philippines. A *barangay* is the smallest political-administrative unit of the Philippine State. Sixteen highly urbanized cities make up Metro Manila. As of 2015, the total population of Metro Manila was 12,877,253. Quezon City itself had a population of 2,936,116, making it the biggest city in Metro Manila and the entire country in terms of population (Fig. 8.1).

Barangay UP Campus was named after the University of the Philippines (UP) campus in Diliman, Quezon City as it is located within a 493-hectare property owned by the university. Originally, a section of the said property was designated as a residential area for the University's employees. Since the early 1970s, however, residents who are no longer connected with the University have continued to live illegally on

Fig. 8.1 Map of Quezon City within Metro Manila

the property while new migrants from the provinces also settled into the area. On June 25, 1975, the Local Government of Quezon City enacted Executive Order No. 24 establishing the said area as "Barangay UP Campus." This law resulted in the creation of two sets of governance structures within the same location: the political structure headed by the *barangay* government of UP Campus and the administrative structure of the University of the Philippines. The *barangay* government is made up of a Barangay Council, headed by a Barangay Captain. All of these *barangay* officials are elected to office by the registered voters coming from, and residents of, Barangay UP Campus. None of these officials are employed by the university. On the other hand, University of the Philippines in Diliman is headed by a Chancellor who is appointed by the university's Board of Regents. There have been many instances in the past where these two sets of governance structures conflicted with one another. For example, the *barangay* government would give permits for the operation of retail stores in the area without securing the permission from University authorities. On the other hand, there would be instances where the University would file notices of eviction to illegally installed structures within the *barangay*, which were tolerated by the Barangay government.

As of 2010, Barangay UP Campus had 5,186 households with a total population of 36,486. Based on a census conducted by the University, only 1,143 of these households live in housing units owned by the university, while the remainder lives in houses illegally built by informal settlers (UP Diliman Office of Community Relations 2011). By 2015, the *barangay*'s population had reached 45,520, representing a growth rate faster than that of Quezon City (Philippine Statistics Authority 2015). The average household consists of 4.48 persons. Unfortunately, there is no updated census of household structures to discern how many are owned and recognized by the university and how many were built illegally. Table 8.1. shows the growth in the population of Barangay UP Campus from 1990 to 2015.

Aside from the academic buildings of UP Diliman, there are university community facilities that are accessible for UP faculty, staff, and students as well as for *barangay* residents. These include two materials recovery facilities, operated by the University's Task Force on Solid Waste Management; the University Health Service, a 25-bed primary health hospital; and the UP Police Station. Barangay UP Campus also owns and operates the following public facilities: Barangay Health Station; Barangay Multi-purpose Covered Court; Barangay Palaris Basketball Court; Barangay Public

Table 8.1 Population of Barangay UP campus, 1990–2015

Year	Population
1990	22,722
1995	25,732
2000	23,226
2010	36,486
2015	45,520

Source Philippine Statistics Authority (2015)

Library; the Centralized Waste Segregation Center; and the Family and Community Healing Center. In addition to the university and *barangay* facilities, two Christian churches (one Roman Catholic and one Protestant) are located in the community. There is no mosque in the *barangay*.

The Research

This case study was conducted primarily by the author with the assistance of the Office of the Vice-Chancellor for Community Affairs (OVCCA) of the University of the Philippines Diliman. The objective of the study was to determine how the Sustainable Development Goals (SDGs) are localized within a community setting.

Before conducting the field research, it was necessary to coordinate with the local government units of Quezon City and Barangay UP Campus. The purpose of the research was explained to the city and *barangay* officials to secure their consent and suggestions. The research methodology included desktop research on the policy instruments used by the Philippine government to attain the SDGs, collection of the *barangay*'s baseline socio-economic data, and workshops with various community stakeholders.

Three separate community workshops were undertaken with three distinct stakeholder groups in Barangay UP Campus, namely: (1) adult local residents; (2) children, and youth; and (3) local transportation stakeholders, including *jeepney* drivers and commuters. In each workshop's opening discussion, we introduced the SDGs, as participants were not yet aware of them. The workshop participants were organized into smaller breakout groups that were given the task of identifying the priority SDGs for their particular sector. The participants were also asked to identify the Filipino values that could be utilized to facilitate the attainment of the SDGs. The following themes and questions were therefore deliberated:

1. Based on the situation and needs in the Philippines, how should the Sustainable Development Goals be aligned with domestic priorities? Participants were encouraged to add and rank additional goals that they deemed relevant or were currently missing.
2. What does it mean to be a Filipino? Participants were asked to provide a list of keywords and explain such things as values, customs, lifestyles, and social norms.
3. How do different Sustainable Development Goals relate to daily life? Participants were encouraged to discuss such topics as purchasing decisions, use of energy and water, lifestyle, budgets, and values.

A separate workshop focusing on children and youth was unique in that it involved interactive and creative activities. The young participants were asked either to draw posters depicting one particular SDG of their choice or draw a logo depicting these SDGs (Fig. 8.2).

Fig. 8.2 Workshop for children and youth at Barangay UP campus

Identifying Local Challenges

The research was able to identify the prevailing and pressing problems within Barangay UP Campus. We learned that 72.8% of the residents are not connected in any way with the University of the Philippines, i.e., they are not regular or contractual employees of UP. From the UP administration perspective, they are "illegal squatters" residing on the university's property. Thus, security of tenure was identified as the number one problem of *barangay* residents, who fear that they may be evicted from their residences at any time.

Moreover, many *barangay* residents do not have access to basic social services such as water and electricity. It was discovered that only 29.2% of the *barangay* households had their own water meters (of Manila Water, the private water service provider for Quezon City). Over one-third of households (36.7%) shared water meters, meaning that water is not available for each household, but two or more households share the same source of tap water. The remaining 25.2%[1] of households had no water meter at all, either buying bottled water or fetching water from unsafe sources, such as water pumps or springs that may be contaminated. Thus, SDG 6 on access to water and sanitation for all is a pressing problem in the community.

[1] The water situation with 8.9% of the households is unknown because the houses were abandoned during the time of the census, and thus no responses were received.

With regards to electricity, it was revealed that 6.2% of the households in Barangay UP Campus do not have electric connection meters. This means that they do not have access to Meralco, the electricity provider for Metro Manila. They may be using candles and kerosene lamps for lighting purposes at night, and this may be a fire hazard. Some of these households connect illegally to existing electric lines and posts, again causing a fire hazard due to potential electrical current overload. These residents are wanting in terms of SDG 7 on access to affordable, reliable, sustainable, and modern energy for all.

There are no elementary and secondary high schools in Barangay UP Campus. While the university has a laboratory school in the area, the UP Integrated School, it is mainly intended for the children of UP employees. The few slots that are available for children of non-UP employees are open to all Metro Manila residents, so *barangay* residents have to compete with other Metro Manila residents to be able to get slots for their children. Usually, Barangay UP Campus residents send their children to the elementary schools located in adjacent *barangays*, such as the Barangay Krus na Ligas Elementary School and the Barangay San Vicente Elementary School. Barangay Captain Isabelita Gravides has submitted a proposal to the UP Diliman Chancellor for the establishment of an elementary school in Barangay UP Campus. However, this proposal has not been acted upon by the university authorities. Thus, it may be said that SDG 4 on inclusive and quality education for all has yet to be attained for Barangay UP Campus.

Concerning health and medical needs, *barangay* residents are allowed to use the University Health Service, but only in cases of emergency. Accordingly, if the attending nurse deems that the problem is not an emergency in nature, the patient is then referred to other government hospitals outside the *barangay*, such as the East Avenue Medical Center. Thus, SDG 3 on healthy lives and well-being for all is still a dream for Barangay UP Campus residents.

In terms of transportation, it has already been mentioned that there are passenger *jeepneys* that ply the lanes of Barangay UP Campus. However, many of these do not burn diesel fuel efficiently, thereby causing air pollution. In 2015, the university experimented with the use of electric vehicles, but this was discontinued in 2017 due to a lack of funds to sustain operations.

Barangay UP Campus faces challenges with regards to the peace and order situation in the community. A 10-year study of crime in the area revealed that during 2005–2014, 1,400 crime incidents were reported to the UP Diliman Police (Camhol 2016). This number excludes unreported crimes and thus, the total crime incidence may be higher. The most commonly reported crimes were theft (51.3%), robbery (19.4%), damage to property (11.2%), and physical injury (9.5%). Since June 2016, when Philippine President Rodrigo Duterte declared an "all-out war" against drugs in the entire country, extra-judicial killings in the *barangay* have increased. From July to September 2016 alone, seven suspected drug users were summarily executed by unknown assailants on Barangay UP Campus premises.

Priority SDGs in the View of Local Residents

Initially, there were differences of opinion among local residents as to which SDGs should be ranked as the highest priority concerns for the community. For example, one breakout group in the workshop for adult residents identified SDG 4 (Quality education), SDG 3 (Good health and well-being), SDG 5 (Gender equality), and SDG 13 (Climate action) as the top priorities for their *barangay*. On the other hand, another group identified the following SDGs as priorities for the *barangay*: SDG 12 (Responsible consumption and production), SDG 13 (Climate action), and SDG 16 (Peace, justice and strong institutions). The results of the discussions for each breakout group were reported in the workshop's plenary session. During this session, there were heated discussions again. Group members were asked why they believed such SDGs were the priority for the *barangay* and why they did not give much weight to other SDGs. After a lengthy plenary discussion, the workshop participants came to a common decision.

This process of deliberations that emerged in the workshop for adult residents was repeated in the two remaining workshops, namely, the workshop for children and youth, and the workshop for transportation stakeholders. The final outcome of all three workshops was an agreement that Barangay UP Campus would give priority to four SDGs. These were, in ranked order: No poverty (SDG 1); No hunger (SDG 2); Good health and well-being (SDG 3); and Quality education (SDG 4). Although residents believed that all of the SDGs are important, they felt that these four SDGs were most relevant for Barangay UP Campus based on the existing local conditions.

During the three workshop discussions, there were also several Filipino values that participants identified as "facilitating values" in working to attain the SDGs. Among the values that were raised were: *bayanihan* (cooperation), hospitality, *malasakit* (empathy), *maka-kalikasan* (environmentally friendly), *matatag* (resilient), *pagtitipid* (prudence), and *utang na loob* (debt of gratitude). After lengthy discussions during plenary sessions, the workshop participants identified three Filipino values that could be best utilized as facilitating mechanisms to attain the SDGs. These are the values of *bayanihan* (cooperation), *pagtitipid* (prudence), and *malasakit* (empathy).

According to Ang (1979, 91), "*Bayanihan* is an indigenous Filipino trait… As anthropologists stress, *bayanihan* is a practical response to both individual and community needs which, under certain circumstances, would be difficult to achieve if people with meagre means did not organize themselves and pool together their resources." On the other hand, *pagtitipid* is defined as "the careful management of financial affairs and frugality in expenditures. It is the wise spending of money and proper use of resources" (Andres 1998, 137). Dagmang equated *malasakit* with the concept of compassion and argues that it informs the helping and help-seeking behavior of Filipinos (2008, 21). The Philippine Development Program (PDP) for 2017–2022 incorporates *malasakit* as one of its strategies. According to the said document, "Under the PDP framework's pillar of Malasakit, the government aims to establish a high-trust society by promoting the Philippine culture and values and nurturing a culture-sensitive development" (NEDA 2017, 12).

During the workshop for children and youth, the participants highlighted the need to develop a learning module on the SDGs that would be part of basic education curricula. Moreover, the participants emphasized that this learning module should be contextualized within the Philippine setting.

Action

Based on the workshop results, an informational poster was developed using the symbol of *Luntiang Puso* (Green Heart) to illustrate how the Filipino values of *bayanihan*, *pagtitipid*, and *malasakit* could facilitate the attainment of the 17 SDGs. This poster (Fig. 8.3.) was produced by the Philippine government's Department of Environment and Natural Resources (DENR) and distributed throughout the country. The logo for Luntiang Puso was designed by one of the participants in the workshop for children and youth.

The workshop participants agreed that the poster, as well as future informational materials, would create greater awareness within the community about the various SDGs, how they are interrelated with one another, and how they might impact their daily lives. It was suggested that future materials be translated into the Filipino language so that they could be better understood.

Another output based on the workshop has been the development of a sample learning module on the SDGs,[2] together with a teacher's guide, aimed for basic level education. This learning module has been pre-tested with elementary and high school students and has been approved by three government agencies, namely the Department of Environment and Natural Resources (DENR), the Department of Education (DepEd), and the Technical Education and Skills Development Authority (TESDA), with funding from the European Union (Castro et al. 2018). Because it has been approved by the national government, it is now expected to be used as a supplementary teaching material in all elementary and high schools in the entire country.

This SDG learning module is intended as a supplementary resource material within the existing elementary and high school level social studies curriculum and will be translated into major Philippine languages. The workshop participants from Barangay UP Campus also suggested that a separate learning module on the SDGs be developed for out-of-school youth in the community. At present, the proposed learning module for out-of-school youth has not yet been developed.

---

[2] The structure of the Learning Module follows the questions raised by the youth participants, such as, on SDG 1 End Poverty: Is my family poor? Are my neighbors considered poor? Who do we consider as poor? Why are so many Filipinos living in poverty? What can I do about the problem of poverty? There are no ready-made answers to the questions, the Teacher-Facilitator will facilitate receiving responses and having class discussions. The draft Learning Module was pre-tested with elementary and high school students coming from selected schools within Metro Manila (not just from Barangay UP Campus). Based on this pre-testing, the final document was approved by the government.

Fig. 8.3 Poster on SDGs based on the output the children and youth workshop

Discussion

Based on the results of this case study, we can see that it is indeed possible to localize research and education about the SDGs within vulnerable communities. Such efforts, in turn, lead to localized strategies toward achieving the SDGs. In this project, local community members themselves identified which among the 17 SDGs are most important for them based on their situations and needs. Each SDG was contextualized based on the particularity of the community. For a poor community such as Barangay UP Campus, the goal of ending poverty is of utmost relevance. According to Maria, an 11-year-old workshop participant, "There will be fewer crimes, maybe no crime at all, in our community if the people are not poor. Young people will go to school because they won't have to work for their families."

In addition, the project confirmed that community members more easily identify with cultural values rather than technical jargon. Therefore, the use of such values can be an effective strategy in achieving the SDGs. It is critical to link global issues with the issues and concerns that are most relevant to the local population. In Barangay UP Campus, the residents identified the values of *bayanihan* (cooperation), *pagtitipid* (prudence), and *malasakit* (empathy) as the most helpful in achieving the four SDGs that community members considered most important. They emphasized that members of the community should work together as one, consistent with the spirit of *bayanihan*, in solving the community's problems. Community residents should not be wasteful of their resources by practicing *pagtitipid*. They also stressed the need to have *malasakit* toward those most in need, such as the poor, the elderly, persons with disabilities, single parents, and other vulnerable groups in the community.

It was only in 2018 that the Barangay UP Campus community identified these three values of *bayanihan*, *pagtitipid*, and *malasakit* as the Filipino values most capable of facilitating the Sustainable Development Goals. Thus it is still too early to say whether these values have in fact contributed to the attainment of the SDGs prioritized by this community. However, we do know that Barangay UP Campus has been undertaking community activities that are consistent with the spirit of *bayanihan* and *malasakit* in 2020–2021, due to the Covid-19 pandemic. Because of the government-imposed lockdown of Metro Manila that began in March 2020, most people, except for "essential workers," have not been allowed to go out of their homes and therefore have been unable to earn a living. Most people have had to rely on food rations coming from the government, but these rations are very limited. Thus, Barangay UP Campus residents established a "community pantry" run by volunteers, where those in need may pick up food supplies based on their respective household's daily needs.

Ideally, the application of the three Filipino values may be institutionalized through their incorporation at basic level education and not just because they are taught by the elders to the younger generation. Still, it may take one or more generations to see if there was indeed a change in the people's behavior.

Finally, because this research project was implemented by the University of the Philippines, we were able to show that educational institutions have the opportunity to facilitate the localization of the SDGs. Universities and other higher learning institutions may extend their outreach programs to neighboring communities. Each college or department of the university may adopt a particular community so that they are also able to monitor progress toward the attainment of SDGs at the local level. In the Philippines, these outreach programs may be undertaken through the National Service Training Program (NSTP), a mandatory program for all college and university students.

Appendix: Philippine Policy Framework On SDGS

The Philippines' strategy in attaining the SDGs is spelled out in the document *AmBisyon Natin 2040: A Long-Term Vision for the Philippines* (NEDA 2016). AmBisyon Natin 2040 was initiated by the National Economic Development Authority based on a visioning process that started in 2015 and was approved by President Duterte by virtue of Executive Order No. 5 in October 2016. The vision document answers the question, "Where do we want to be?" The vision for the Philippines for the next 25 years is for the country to achieve "*Matatag, Maginhawa, at Panatag na Buhay.*" This means that Filipinos are strongly rooted (*matatag*), comfortable (*maginhawa*), and secure (*panatag*). Concretely, by 2040, the Philippines is envisioned to be: (1) A prosperous middle-class society where no one is poor; (2) People live long and healthy lives and are smart and innovative; and (3) The country is a high-trust society where families thrive in vibrant, culturally diverse, and resilient communities (NEDA 2016). Education is mentioned as a key element in improving the quality of life by enabling students to acquire the foundational literacies (reading, numeracy, scientific literacy, ICT literacy, economic and financial literacy, and cultural and civic literacy); obtain competencies (in critical thinking, problem-solving, creativity, communication, and collaboration); and develop character qualities (such as curiosity, innovativeness, persistence and grit, adaptability, leadership, and social and cultural awareness).

Another important government document is *The Philippine Development Plan, 2017–2022* (NEDA 2017). This is the Philippine government's five-year development plan guided by AmBisyon Natin 2040. It builds upon what has already been accomplished in the previous development plan, the Philippine Development Plan, 2011–2016. The PDP 2017–2022 has the following targets:

1. The Philippines will be an upper middle-income country by 2022.
2. Growth will be more inclusive as manifested by a lower poverty incidence in rural areas, from 30% in 2015 to 20% in 2022.
3. The Philippines will have a high level of development by 2022.
4. The unemployment rate will decline from the current 5.5% to 3–5% in 2022.
5. There will be greater trust in government and society.
6. Individuals and communities will be more resilient.
7. Filipinos will have greater drive for innovation.

To be able to achieve the aforementioned objectives, the PDP identified the following strategies:

1. Enhancing the social fabric: The aim is to regain people's trust in public institutions and cultivate trust in fellow Filipinos.
2. Inequality-reducing transformation: Ordinary Filipinos will feel the "*Pagbabago*" (change).
3. Increasing potential growth: It is imperative that economic growth is accelerated and sustained for "*Patuloy na Pag-unlad*" (continued development).
4. Enabling and supportive economic development.

5. Foundations for sustainable development.
6. Migration and development.

It is interesting to note that there is an entire chapter (Chapter 7) in the PDP that is devoted to promoting Philippine culture and values. To be able to do this, the plan identified four outcomes in the realm of culture, namely: (1) our diverse cultures valued; (2) values for the common good inculcated; (3) *pagkamalikhain* or creative excellence enhanced; and (4) culture-sensitive governance and development strengthened.

References

Andres, T. 1998. *People Empowerment by Filipino Values*. Manila: Rex Book Store.

Ang, G. 1979. "The Bayanihan Spirit: Dead or Alive?" *Philippine Quarterly of Culture and Society* 7: 91–105.

Camhol, A. 2016. *Who Gets Whacked, When, and Where: A Spatio-temporal Analysis of Crime in the UP Diliman Campus*. Quezon City: University of the Philippines.

Castro, N., R. Chanandran, and C. Gunawardena. 2018. *Mga Hangarin para sa Likas-kayang Pag-unlad* (Teachers' Guide and Sample Learning Module). Quezon City: Department of Environment and Natural Resources.

Dagmang, F. 2008. "Malasakit (Compassion) and Damay (Empathetic Assistance): An Indigenous World-Embedded Compensation and Resistance at the Margins." *Unitas* 81, no. 2: 307–336.

NEDA (National Economic Development Authority). 2016. *AmBisyon Natin 2040: A Long-Term Vision for the Philippines*. Pasig City: NEDA.

NEDA (National Economic Development Authority). 2017. *Philippine Development Plan, 2017–2022*. Pasig City: NEDA.

Philippine Statistics Authority. 2015. *Census of Population and Housing, 2015*. Manila: PSA.

UP Diliman Office of Community Relations. 2011. *Census of Self-Built Units*. Quezon City: University of the Philippines Diliman.

Chapter 9
Sustaining Biogas Production in the Indonesian Province of West Nusa Tenggara

Niken Arumdati

Highlights

- Biogas is proliferating on the energy-challenged island of Lombok, despite challenges and setbacks.
- Biogas sustainability is a result of collaborative work and collaborative learning on the part of various parties and sectors; its success requires ensuring that all their voices are heard, that each sector contributes according to its capacities, and that plans and designs can be flexible enough to adapt to changing conditions and new possibilities.
- Assessing biogas production against five dimensions of sustainability—technical, economic, social, environmental, and institutional—allows us to see more clearly both what needs to be done to make this form of renewable energy a sustainable option, and the many ways it contributes to sustainability, especially of society and the environment.
- In the midst of Indonesia's drive to expand renewable energy, this chapter provides a rare glimpse into the learning journey of a seasoned government officer as she worked with diverse stakeholders on a renewable energy project in a single province.

Introduction

Like much of eastern Indonesia, West Nusa Tenggara Province's 864 idyllic tropical islands—of which 461 are inhabited—all face the challenge of poor access to energy. Of the two largest islands, Lombok and Sumbawa, Lombok is the one that has become widely known as a desirable tourist destination, receiving more than one million

N. Arumdati (✉)
Head of the Renewable Energy Development Subdivision of the Office for Energy and Mineral Resources of the Provincial Government of West Nusa Tenggara, West Nusa Tenggara, Indonesia

M. Indrawan et al. (eds.), *Civic Engagement in Asia*,
https://doi.org/10.1007/978-981-16-9384-7_9

visitors each year. Yet, limited electricity access constrains the people's livelihoods and the local economy. Though the potential in this province for mini-hydro alone could reach 100 MW (Muchtar et al. 2014), only a little more than 41 MW installed capacity of renewables has been made available so far.

As Head of the Renewable Energy Development Subdivision of the Office for Energy and Mineral Resources for West Nusa Tenggara, a subdivision with only three staff members, I am the person most directly responsible for the development of renewable energy in the province. In this chapter, I reflect on our experience of developing small-scale biogas plants as part of the renewable energy plan for the province, looking especially at the different dimensions of what makes such projects sustainable. Beyond that, I consider as well the learning process that our decisions and actions have put in motion. This learning has been multi-directional, as it is not only we who instruct the public and encourage governments to invest in new technologies; it has also been successful and entrepreneurial local users of biogas, as well as supportive NGOs, who have instructed us and worked with us to explore what is possible.

Our attempts to expand renewable energy in the province have engaged us with members of the public in various ways. For example, the need for monitoring and evaluation of government-supported projects such as biogas brings us into contact with local farmers and ranchers. When I visited local areas, it was especially the women with whom I would interact, because the men were often out working in the fields. These interactions required by government monitoring became an important vehicle for mutual learning. Furthermore, our interactions with NGOs or small workshops and entrepreneurs have often revolved around solving problems that only appeared once we had set a project in motion. Such engagements have also been an important site of learning. Finally, the COVID-19 pandemic likewise presented new challenges that we could address only through discussion and sharing our expertise and bodies of knowledge with the aim of finding new solutions.

In this chapter, I describe some of the challenges we encountered in developing biogas in the province, and how we have adapted to the conditions of governing as well as to conditions on the ground, including the restrictions introduced by the COVID-19 pandemic. Our aim has been to continue to expand biogas as a viable renewable energy solution for the province, despite setbacks. This story is one example of how policymakers too can undergo an education that changes our perspectives and helps us to see things we did not see before. Finally, I write with the hope that some of what we have learned may be useful for others who are developing or would like to develop biogas in their regions.

Background to Biogas Development in West Nusa Tenggara

When the newly appointed governor of the Indonesian Province of West Nusa Tenggara began his tenure in 2009, one of his primary programs was to increase the number of livestock raised in the province by a figure of 1,000,000 during his term in office. Meanwhile, a livestock-related environmental concern was already arising. Farmers were disposing of manure in rivers and other waterways, a practice that was contaminating waters and affecting human and ecosystem health.

As the provincial level government agency responsible for encouraging the utilization of renewable energy, the Office for Energy and Mineral Resources of West Nusa Tenggara province (henceforth Dinas ESDM) had been developing biogas projects since 2007. However, most of the projects were developed only as pilot projects; at that time, the price of kerosene was still heavily subsidized by the central government, and communities were reluctant to use biogas as a substitute fuel for cooking.

In 2009, the central government planned to replace kerosene with subsidized liquid petroleum gas (known as "LPG 3 kg"). However, distribution was problematic due to lack of access in rural or remote areas and to the limited supply of the liquid petroleum gas itself. To fill this gap and reduce the environmental impact of livestock, biogas appeared to be a likely solution. This is because the installation cost is not too high, animal dung as feedstock for biogas is easily available, and small-scale biogas technology has been successful in a number of Asian countries.

Biogas plants convert animal dung and various other organic materials into combustible methane gas through a process of controlled fermentation. Household-scale biogas can produce enough energy to meet the cooking needs of a single household in a day. And it has a wide range of uses, from simple gas stoves for cooking to lamps for lighting. On average, farmers with at least two cows can generate sufficient biogas to meet their daily basic cooking and lighting needs. While our initial aim was to develop biogas as a replacement fuel for cooking and lighting, our focus shifted over time to another of its important benefits: the bio-slurry left over from this process is easily collected and can be used as organic fertilizer to improve crop yields. Through its use as a fertilizer and through other means that we learned about over time, the bio-slurry could thus be a source of income generation as well.

The promotion of biogas is believed to have contributed positively to the Sustainable Development Goals (SDGs), especially SDG 7 (Affordable and Clean Energy) and 8 (Decent Work and Economic Growth). Biogas plants produce renewable energy, which can be an alternative or substitute for fossil fuels previously used by the beneficiaries (SDG7). Furthermore, biogas also plays a role in achieving renewable energy targets in national energy policy and in provincial government plans for increasing energy supplies at the regional level. In addition, this program encourages job creation, utilization of bio-slurry, and local food security, while reducing household expenses (SDG 8). Yet, there remains a long way to go to realizing all of these potentials of biogas to their fullest.

Discoveries About the Technical Sustainability of Biogas Plants in West Nusa Tenggara

In a 2008 article, Elisabeth Ilskog proposed five dimensions through which the sustainability of rural electrification projects could be assessed: technical, economic, social/ethical, environmental, and institutional. I apply Ilskog's scheme to assessing the sustainability of biogas energy in West Nusa Tenggara province, beginning with technical sustainability. Technical sustainability refers to the sustainability of material and support services that ensure the operation of biogas plants, so that they run well and do not face major problems (Wahyudi et al. 2015).

Our efforts to promote the installation of household-scale biogas plants on the two main islands of Lombok and Sumbawa began in earnest in 2012. As we entered the phase of monitoring and evaluation of biogas installation, our attention to the plants' technical sustainability required looking at two main areas: the regular availability of feedstock for the plants and the availability of servicing. As we will see, we ran into difficulties in both areas, leading to new assessments and conclusions.[1]

As manure is the essential feedstock for the biogas plants in our province, the availability of manure is a necessary requirement for the successful operation of the plants. Based on data obtained from Dinas ESDM, this table first presents the figures for the actual cattle population in ten regencies and municipalities on Lombok and Sumbawa in 2011, then the estimated production of manure given the number of cattle, and finally the expected biogas production given available quantities of manure. The first five regencies and municipalities are on Lombok, while the second five are on Sumbawa.

	Regencies/ Municipalities	Number of Cattle (Cows and Buffalos)	Manure Production (kg/day)	Biogas Production (m^3/day)
1	Kota Mataram	1,322	19,830	714
2	Lombok Barat	81,155	1,217,325	43,824
3	Lombok Utara	66,165	992,475	35,729
4	Lombok Tengah	112,058	1,680,870	60,511
5	Lombok Timur	86,022	1,290,330	46,451
6	Sumbawa Barat	54,621	819,315	29,495
7	Sumbawa	211,332	3,169,980	114,119
8	Dompu	92,288	1,384,320	49,835
9	Kota Bima	18,951	284,265	10,234
10	Bima	127,941	1,919,115	69,089

(continued)

[1] Apart from feedstock, water is the second most important component of the biogas plants' operations. Biogas plants built in water-rich areas such as the northern part of Lombok Island do not face water constraints. However, the biogas plants built in the southern part of Lombok Island or Sumbawa Island are often constrained by lack of water, especially during the dry season.

(continued)

	Regencies/ Municipalities	Number of Cattle (Cows and Buffalos)	Manure Production (kg/day)	Biogas Production (m^3/day)
Total		851,855	12,777,825	460,002

Source Central Bureau of Statistics, 2011

Judging simply from the table, the technical sustainability of biogas plants on Lombok and Sumbawa should have been very comparable. The number of cattle in the various municipalities and regencies of both islands was sufficient to supply feedstock to the biogas plants on both islands.[2] And farmers and ranchers from both of these islands were enthusiastic about installing the biogas plants and submitted proposals to our office for support to do so. The sites looked promising since the number of cattle were high, and the people were quite interested. However, within only a few months, we had learned through monitoring and evaluation trips that most of the biogas plants on Sumbawa were no longer in operation. Why? Women of the household would tell me that it was far too much effort to collect the manure. In fact, the methods of the farmer-ranchers on the two islands are vastly different. On the island of Sumbawa, grassland is still widely available, therefore, farmers do not pen their cattle, but let them graze freely on the grassland. This of course means the manure would be widely scattered. Meanwhile, on the island of Lombok, because many grasslands have been converted into residential areas, cattle are put into stalls by farmers, either at their homes or in shared or communal stalls. As a result, it is much easier for farmers to collect manure as feedstock for biogas plants on Lombok than it is on Sumbawa.

At that time, it would have made sense to initiate a pause in the construction of new biogas plants on Sumbawa, given what we had learned. Unfortunately, we were under political pressure to continue to approve new plants on both Lombok and Sumbawa, as the central government funds come through the provincial government, and we were bound within the yearly plan to put equal resources for such projects into both islands. This was one instance in which the conditions of financing, governance, and continued support put limitations on our decision-making. We would be free to change plans once we were into a new annual cycle. But due to political pressures we were committed to an approach that would likely yield a high rate of failure for the Sumbawa plants through the rest of 2012. At the same time, we knew that we would have to be evaluated by the central government, so we did our best to limit new installations of the plants in Sumbawa, which we had by then concluded were a risky investment.

We also faced difficulties when it came to providing sustainable servicing of the biogas plants. The construction of the biogas plants built on Lombok Island has been carried out by Construction Partner Organizations (CPOs) and supervisors, who

[2] The cattle population on both islands has continued to increase every year. This is partly due to programs run by the provincial level governments, especially the program called "The Village of One Thousand Cows," initiated in 2012.

have trained and certified masons to build biogas plants according to the Indonesian National Standard (SNI). The CPOs were actually formed as private enterprises on the initiative of Hivos, a Netherlands-based international NGO that has played a key role in supporting the installation of biogas plants in many parts of Indonesia. In addition to working on construction, CPOs also provide biogas spare parts and appliances. However, because the distribution of the biogas plants has been widely spread, the repair facilities and shops selling spare parts are centralized in the cities of Lombok, including the provincial capital of Mataram. This makes it difficult for users who live far from these cities to access these services. The problem is much more difficult for farmers in Sumbawa, who would require 10–12 hours of travel to reach servicing centers in Lombok. To solve this problem, CPOs will need to establish servicing networks with groups of biogas farmers in each location where the biogas plants are built, so that in each village there would be a distribution chain for biogas spare parts and appliances. This conclusion is supported by Bhattacharyya (2012), who writes that the future of biogas technology depends on developing adequate servicing networks.

What We Have Learned About Making Biogas Plants Sustainable Economically

Financial Support

Most of the biogas plants built on the island of Lombok in particular, and Indonesia in general, are fully funded by the government or partially subsidized by NGOs, especially Hivos, in collaboration with the Indonesian civil society organization Yayasan Rumah Energi (YRE). Both the provincial government and the central government allocate funds annually to build biogas plants in order to pursue the renewable energy contribution target of 23% in the national energy policy. However, since 2018, the allocation for biogas development has progressively decreased, because most of the government's budget has been focused on improving infrastructure after the big earthquake in 2018. Furthermore, since 2020 an even bigger budget has been needed for economic recovery during the COVID-19 pandemic and for health infrastructure and programs. Since the construction of biogas plants is still highly dependent on government funds and NGO subsidies, the number of biogas installations on Lombok Island has been decreasing sharply in recent years.

In order to encourage farmers' interest in installing biogas as independent investors, since 2013 financing has been introduced through Kiva. Kiva is an international non-profit organization based in San Francisco that provides soft loans to farmers with zero interest, requiring users to pay back the loans in several installments within a specific period of time (Hivos 2014b). However, my contacts in Hivos told me that many of the farmers who took out loans failed to pay back their debts.

The reasons, among others, include that there are still biogas plants being built with full subsidies from the government.

Furthermore, in talking to farmers, I learned that many of them feel that the amount of money deposited into Kiva in installments each month is too large compared to the benefits received. Below I discuss how our expanded view of the benefits of biogas could shift their calculations.

First, however, I would like to share how we made adjustments to the design of the biogas plants, in collaboration with our NGO partner, Hivos. Our new design was partly in response to the fact that government funding during the pandemic was drastically reduced and had become more unpredictable. Under those conditions, we wanted to make it easier economically for farmers to invest their own money in building biogas plants. During the past few years, we have had a total of only about 100 households in Lombok invest their own money in building a biogas plant. However, our decision was also in response to the fact that Hivos's earlier design of 4–6 cubic meter plants required a larger installation space, 3.7 square meters, which was often not available in urban or near-urban settings. The new design of 2 cubic meters required a smaller installation area and was much more economical to pay for. The larger earlier design required a total investment of 14–15 million IDR (roughly $1,000 USD), while the new smaller size required only 5 million IDR (about $350 USD). And Hivos offered a subsidy of 3 million IDR ($210 USD), leaving only IDR 2 million ($140 USD) for the individual farmer to invest. Comparing figures from 2020 and 2021, our new design has been very successful in encouraging user investment in building the plants. For all of 2020, we had 20 households invest their own money in building a biogas plant. At the time of writing, mid-June of 2021, we already have 30 households that have invested their own money, using the new smaller design. This is a very good step in the right direction.

Using Biogas to Reduce Household Expenses and to Generate Income

Initially, we were motivated to promote biogas primarily as a renewable alternative for home cooking and lighting needs, and in doing so to help meet renewable energy targets. Since the introduction of LPG in 2009 and the initiation of a national program for it to replace the use of subsidized kerosene in the household sector, LPG has become the main source of fuel for domestic cooking in Indonesia, including on Lombok Island.[3] We realized that farmers who use biogas plants would also benefit from the savings of not having to buy LPG each month. Currently, LPG prices at the retail level range from IDR 17,000 to IDR 18,000 ($1.18 USD–$1.25 USD) per 3 kg. However, distribution problems sometimes occur so that LPG becomes scarce.

[3] It remains true that in rural areas, many people still do depend on traditional biomass such as fuelwood for cooking, because the raw material is normally widely available around them, and it is free.

When this happens, in rural areas far from the main LPG supply chain, the price can be as high as IDR 22,000 ($1.53 USD) per 3 kg. Based on data released by the Ministry of Energy and Mineral Resources, the average consumption of LPG for poor households is three 3 kg LPG cylinders per month, so the potential savings per month with biogas can reach IDR 66,000 ($4.60 USD) per month.

Expanding the use of renewable energy was of course our key mandate within Dinas ESDM. Yet, our biogas users played a big role in driving home to me personally the potentials for generating additional household income from the biogas plants. One outstanding example was from 2018, when I learned of a very enterprising woman in West Lombok by the name of Umi Mingsih, who was the wife of a village headman. She ran a recycling bank for plastics and other waste, but she was also using the bio-slurry, a byproduct of the biogas plants, as a nutrient for her fisheries. The manure used as feedstock in the biogas plants needs to be mixed with water, and the byproduct of producing the gas is this bio-slurry. She would dry it first, and then add it to her fishponds. She also used the bio-slurry as an organic fertilizer for her garden. A local university did a study and found that the bio-slurry contained significant amounts of nutrients. Encounters such as the one I had with Umi Mingsih had a strong impact on me. Combined with Hivos's commitment to exploring all the benefits of biogas, and their decision to initiate training in the use of bio-slurry as an organic fertilizer, discussions took place that brought about a shift in our messaging as well, a process that is still underway today.

The bio-slurry from biogas plants can of course also be sold as organic fertilizer, and this brings in additional household income. On the island of Lombok, the price of solid bio-slurry is IDR 1,000 ($.07 USD) per kg. This bio-slurry is only used in the local farming community, though. It is not for sale on a large scale, because if bio-slurry is traded on a large scale, a series of laboratory and field trials is needed to determine the nutrient content and dosage of fertilizer application. In other words, at the moment there is an economic contribution obtained from the sale of bio-slurry, but the economic impact that is felt is still small. Nevertheless, I believe that such economic benefits could help generate interest once again on the island of Sumbawa, though the farmers' learning process will also take time. And our messaging needs to shift yet further in that direction.

Using Biogas to Generate Employment

Based on a study conducted by Mwakaje (2008), the adoption of biogas technology has the potential to create employment opportunities, both in local districts and in countrywide. This can happen if the construction of the biogas plants does not rely too much on government funds every year. Biogas funding from the government is uncertain because budget alignments depend on the political will of regional leaders and on which work programs are considered a priority. However, if biogas is able to be financed from existing demand with minor government intervention, the supporting components such as contractors, masons, training institutions, and input suppliers will grow. Under current conditions, the role of the government, which has been too dominant in terms of funding, results in uncertainty. This uncertainty makes it difficult for supporting businesses to grow because of the large investment risks.

Social and Environmental Sustainability

Here, I would like to briefly address additional social and environmental benefits of adopting biogas, before concluding with a discussion of what is needed to promote the institutional sustainability of this form of energy production in West Nusa Tenggara province. In terms of environment and the climate, manure contains nutrients, such as nitrogen, phosphorus, and other minerals, that enable it to reduce and replace the usage of chemical fertilizer (Marañón et al. 2011). According to Amon et al. (2007), the anaerobic digestion (AD) involved in producing biogas is one of the best ways to reduce greenhouse gas emissions compared to other options. AD also provides controlled CH_4 emissions during storage and by burning CH_4. It does release CO_2, though CO_2 has less Global Warming Potential (GWP) than CH_4, and the amounts are smaller than when fossil fuels are burned. The role of chemical fertilizers and pesticides in degrading the soil's ability to sequester carbon is a large and growing subject at the present time, and there is even a recent and popularly accessible film dedicated to this topic, entitled *Kiss the Ground* (Tickell and Tickell 2020).[4] For our purposes, it is perhaps enough to indicate the importance for soils and the climate of reducing or eliminating the use of chemical fertilizers, and replacing them with organic fertilizers, such as the bio-slurry produced in biogas plants.

In terms of social sustainability, biogas plants make important contributions to human health. Manure contains pathogenic bacteria that can be dangerous, a matter of serious concern if the manure is disposed into rivers or waterways, which the local people depend on for their clean water needs. This was in fact one of the important motivations for our subdivision of Dinas ESDM to begin promoting biogas. Apart from the noxious odor of manure disposed in waterways and its impact on aquatic organisms, consuming water contaminated with manure can cause disease in humans. Biogas also contributes to human health by reducing the indoor air pollution that occurs when cooking with firewood. Biogas utilization reduces this risk, and since one of the indicators of a decent standard of living is access to clean energy, it can be said that biogas has been positively contributing to improving users' standard of living (Rao and Pachauri 2017).

On several occasions my supervisor asked me to produce short video testimonials on the benefits enjoyed by users of biogas. These videotaped interviews were also an education for me. Women reported to me that they were very pleased that once they began using biogas as a cooking fuel, and they did not have to spend so much time and go to such great lengths to gather firewood to feed their stoves. In rural

[4] The film *Kiss the Ground* is an award-winning documentary film by director-producers Joshua and Rebecca Tickell, based on the book by Josh Tickell (2018).

households, there is a division of roles in which women are usually responsible for cooking activities, including providing fuelwoods for cooking. Using biogas plants will reduce the time spent gathering fuelwood and cooking so that women will have time for productive activities, education, and involvement in community activities. Another point to note is that many people feel insecure when using LPG in Indonesia due to accidents reported in the mass media (Budya and Arofat 2011). Compared to LPG, biogas is safer for woman working in the kitchen because of the standardized system and low pressure of the biogas (Hivos 2014a). For these reasons, women will benefit significantly from the use of biogas, and we can say that in this way biogas contributes to gender equality in households that rely on it.

Institutional Sustainability

Ironically, the videos I produced, highlighting the benefits of biogas, were not only to persuade farmers and others to install biogas plants. They were also to persuade changing provincial governments to continue supporting the promotion of this form of renewable energy. In this section, I share my view that while government support has been crucial in growing biogas through what I call "People-Private-Public" or PPP partnerships; we must also encourage individuals and households to invest on their own in biogas plants, in "People-Private" or PP partnerships. This will help to spread the use of biogas and reduce the uncertainties created by shifting government priorities. This issue has become even more critical in the recent period when budgets have shifted to building infrastructure following the 2018 earthquake and to responding to the COVID-19 pandemic.

The three main stakeholders involved in biogas development on Lombok Island are the provincial government, Hivos/YRE, and farmers/users. The role of the provincial government is now quite dominant. Previously, the energy sector was fully managed by the central government. However, since the implementation of regional autonomy, there has been a division of authority between the central government and the provincial government. The provincial government has the authority to formulate policies in the energy sector and implement them.

Based on the Regional Energy General Plan for West Nusa Tenggara Province, there is a target of 23% contribution of renewable energy to the regional energy mix by 2025, and biogas is included in this. Biogas plays an important role in providing alternative energy for cooking that is environmentally friendly and can replace the use of LPG and fuelwood. For this reason, every year the provincial government provides a budget for biogas development, even though the amount of these funds fluctuates every year in accordance with the regional fiscal conditions.

The role of Hivos/YRE is also crucial. Hivos/YRE ensures the sustainability of biogas plants by providing standardized technology, skilled masons, and after-sales service. In developing countries, 50% of biogas plants are non-functional due to lack of maintenance and repair of existing facilities (Bond and Templeton 2011). The budget of the provincial level government, as in West Nusa Tenggara, only provides for biogas construction and does not include the cost of biogas maintenance. In order to provide reliable after-sales service, Hivos sets aside a small portion of the payment it receives for construction costs and uses it for monitoring and maintenance purposes. Monitoring and maintenance are carried out periodically, starting from three years after the biogas is completed.

The following chart provides a graphic view of the different roles that the provincial government, the private and non-profit sector, and users play in PPP partnerships (Fig. 9.1).

Currently, the role of the user is still limited for the most part to being a beneficiary. To support the sustainability of biogas, the role of the user needs to be expanded, in my view. They must be involved in biogas plant development from the planning to the construction and operational stages. At the planning stage, users need to express their opinion about wanting to have access to clean energy in the form of biogas by submitting proposals to the provincial government. After the proposal is submitted, a feasibility study will be performed to confirm the possibility of carrying out the biogas project. At this stage, user engagement can be encouraged by the provincial government through interacting and building good relationships with the potential users, attracting their interest and willingness to participate in the project, and asking for their commitment to look after the biogas plant after commissioning. In addition,

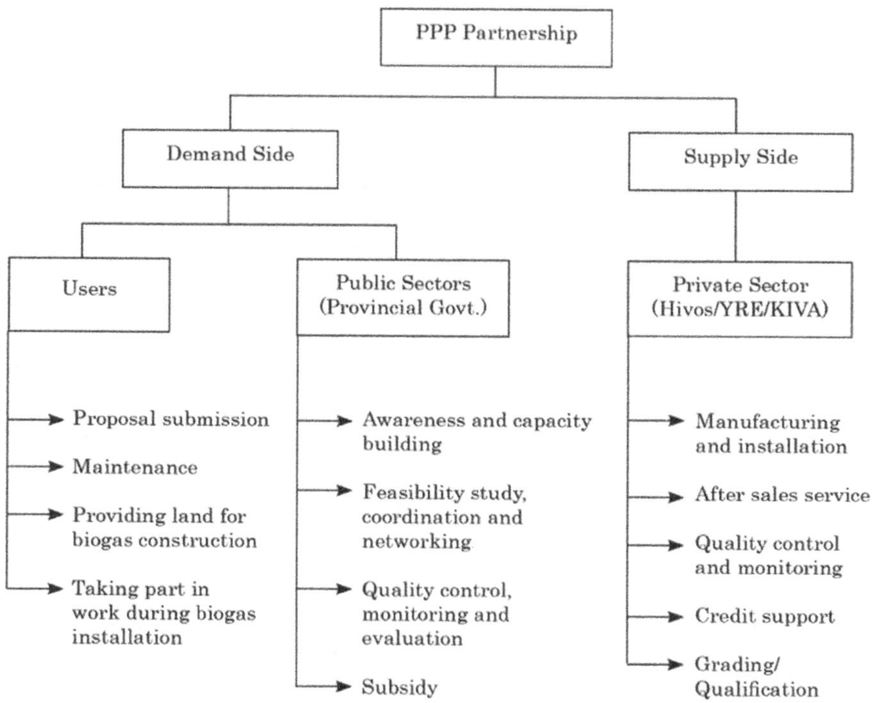

Fig. 9.1 Key stakeholders and their roles in Public Private People (PPP) Partnerships

because this program is user-initiated, it is necessary to emphasize tailor-made policies already in the planning stages. Biogas design takes into consideration flexible and adaptive planning. It is important to know that this design is not rigid; rather, it can be adapted to real situations in the field.

At the construction stage, users should be more engaged in the biogas project. Once the construction work has started, users who may still be unclear as to what a biogas plant would look like could directly witness and learn step-by-step about the biogas plant that is about to be established. Lastly, to improve the capability of users in general not merely in operating biogas, but also in doing simple troubleshooting and utilizing bio-slurry as a biogas byproduct, the training and capacity building for users undertaken by Hivos/YRE is of key importance. Capacity building is intended to be a process for all beneficiaries to become equally independent and to understand how to make the most of the existence of the biogas plant for advancing their lives, and not to see themselves simply as recipients of charity (Afifah 2016).

In essence, I am proposing that more effort and attention be given to promoting PP partnerships, in which local farmers and potential users of biogas plants bear some portion of the investment cost, in collaboration with private non-profit groups such as Hivos and YRE and their affiliated private enterprises. Not only do some households have the financial ability to invest, especially given the subsidies that Hivos provides; in addition, environmental awareness is growing, and people increasingly feel the need to shift to green forms of energy. Finally, disseminating the message that using biogas brings many additional benefits in the way of health, income and employment generation, and reducing women's household work, would likely encourage more households to make the necessary investment. What I am proposing is that in addition to the PPP partnerships depicted in the graph above, we should encourage PP partnerships that do not need to rely so much on the government sector, and that have an expanded and more engaged role for users.

Discussion

International lessons are relevant to our experiences in West Nusa Tenggara. In developing countries, about 50% of biogas plants are not functional due to lack of maintenance and repair of expansion facilities (Bond and Templeton 2011). However, successful ones yield added values. A study in Tanzania showed that farmer households using biogas had a higher average annual income than farmer households that did not use biogas, and this is because of the bio-slurry as side product (Laramee and Davis 2013).

Lessons may also be drawn from our neighboring East Nusa Tenggara province, where a civil society organization—a local NGO called Geng Motor Imut (GMI)—applied locally developed portable biogas digesters and used otherwise foul-smelling pig manure from the neighborhood to generate electricity (Rusdianto 2014). The relatively quick adoption of the local technology by the local households was partly due to the fact that the proponents are well-known as part of the grassroots community. Therefore, this kind of approach offers a social solution that goes beyond technological or engineering matters.

West Nusa Tenggara Province, in particular Lombok Island, has abundant potential for biogas development since it shows a positive trend, and it is fully supported by the provincial level government. Nevertheless, development of biogas has been considered slow because our assessments do not consider indicators in sustainability. Assessment of these indicators can improve government policies going forward, so that in the future the dissemination of biogas plants in Lombok Island can be more rapid, more free from reliance on government or NGO subsidies, and more sustainable.

Taking into account the various dimensions of assessing the sustainability of biogas plants, we can draw the following lessons from our experience of this biogas project. Firstly, biogas development must pay attention to the habits of local farmers: whether cattle are herded on grassland or put into stalls makes a significant difference in whether manure as feedstock can be efficiently collected. Secondly, it is also important to establish servicing networks with groups of biogas farmers in each location where the biogas plants are built, so that in each village there would be a distribution chain for biogas spare parts and appliances. I believe this is possible as the numbers of users grow. Farmer habits and servicing are both aspects of making the biogas plants sustainable in technical terms.

Thirdly, to reduce dependence on government funds because of budget uncertainties, and at the same time to attract more farmers to use biogas plants, a market should be established. A market that is built up around all aspects of biogas plants—financing, construction, maintenance, servicing, byproduct use, skills training, and design adaptation—will also allow all the supporting businesses to grow. Furthermore, the financing of biogas production can now also be supported by Indonesia's increasing village funds. The village funds are in large part accessible by the legal entity known as the village-owned enterprise. However, this will also require developing various skills, as currently people often lack the skills and ideas necessary to grow a sustainable business.

Fourthly, the value of clean energy to health should be highlighted, at both the household and the society level. The use of clean energy such as biogas for cooking has directly led to a decrease in the occurrence of respiratory diseases caused by household air pollution. And every small step toward renewable energy also mitigates the enormous and disruptive impacts to every society of climate change.

Fifthly, although the economic contribution obtained from the sale of bio-slurry is considered small, the utilization of bio-slurry to generate additional income can be another motivator for people to adopt biogas energy in their homes. To this, we can add the potential for new forms of employment as people become involved in the marketing, construction, and servicing of the plants. Finally, as I have argued in the section on strengthening the institutional sustainability of biogas in West Nusa Tenggara, the role of the user needs to be expanded. Users must be involved in biogas plant development from the planning, construction, and operational stages.

Final Reflection

Our small team working on renewable energy for West Nusa Tenggara province has promoted renewable energy in a variety of areas. These include promoting small photovoltaic (solar cell) power plants in villages, as well as mini- and micro-hydropower. With the onset of the pandemic in 2020, the government restricted people's mobility as well as outdoor activities, but in this context people's interest in biking grew rapidly. So we took advantage of this momentum to help our local economy recover by promoting a healthier as well as cleaner energy lifestyle to the public in Mataram. This took the form of partnering with small enterprises in Mataram City to produce electric bikes. We now work with local workshops there who can produce the bikes themselves, our role being to facilitate their access to the market by giving them training, for example, on how to put their product onto eCommerce platforms.

Similarly, the lessons we have drawn from our experience of promoting biogas in West Nusa Tenggara have been possible as a result of a collective learning process. We have learned from the results of our own actions as government officers and policymakers, from our study of cases far afield, from our partners in the private and non-profit sector, and from the users and beneficiaries of the biogas plants themselves. We have been most successful when we have been able to listen to these many voices, and when we have been able to respond flexibly to the ever-changing conditions in our region. The strong response during the pandemic to our promotion of a new biogas plant design by our partner Hivos—a smaller design fit for urban spaces and within financial reach of a larger number of households—is one of the more promising examples of such flexibility and adaptation. The fact that Hivos built a total of 70 new plants on Lombok between January and May of 2021, mostly in urban areas, is also promising. I feel confident that as more people understand the benefits of biogas plants and see more clearly not only what makes them sustainable but also their contribution to sustainability more broadly, we will succeed in our effort to make this form of renewable energy a popular option in our province.

References

Afifah, H. 2016. *Evaluation of State-sponsored Rural Electrification Project in Indonesia: A Case Study of Pengantap Hamlet, West Nusa Tenggara, and Lessons from a Process and Learning-Based Approach.* Master's Thesis. Eindhoven University of Technology.

Amon, T., B. Amon, V. Kryvoruchko, A. Machmüller, K. Hopfner-Sixt, V. Bodiroza, R. Hrbek, et al. 2007. "Methane Production through Anaerobic Digestion of Various Energy Crops Grown in Sustainable Crop Rotations." *Bioresource Technology* 98, no. 17: 3204–3212.

Bhattacharyya, S.C. 2012. "Energy Access Programmes and Sustainable Development: A Critical Review and Analysis." *Energy for Sustainable Development* 16: 260–271.

Bond, T., and M.R. Templeton. 2011. "History and Future of Domestic Biogas Plants in the Developing World." *Energy for Sustainable Development* 15: 347–354.

Budya, H., and M.Y. Arofat. 2011. "Providing Cleaner Energy Access in Indonesia through the Mega Project of Kerosene Conversion to LPG." *Energy Policy* 39: 7575–7586.
Central Bureau of Statistics (BPS). 2020. "Statistics of West Nusa Tenggara."
Hivos. 2014a. "Final Report, Indonesia Domestic Biogas Programme, May 2009–December 2013."
Hivos. 2014b. "Interim Report, Indonesia Domestic Biogas Programme, January–June 2014."
Ilskog, E. 2008. "Indicators for Assessment of Rural Electrification—An Approach for the Comparison of Apples and Pears." *Energy Policy* 36: 2665–2673.
Laramee, J., and J. Davis. 2013. "Economic and Environmental Impacts of Domestic Bio-digesters: Evidence from Arusha, Tanzania." *Energy for Sustainable Development* 17: 296–304.
Marañón, E., A. Salter, L. Castrillón, S. Heaven, and Y. Fernández-Nava. 2011. "Reducing the Environmental Impact of Methane Emissions from Dairy Farms by Anaerobic Digestion of Cattle Waste." *Waste Management* 31: 1745–1751.
Muchtar R., L. Cameron, and M. Indrawan. 2014. "Taking Action to Cut Emissions and Boost Development." Climate and Development Knowledge Network. https://cdkn.org/2014/03/feature-taking-action-to-cut-emissions-and-boostdevelopment/.
Mwakaje, A.G. 2008. "Dairy Farming and Biogas Use in Rungwe District, South-west Tanzania: A Study of Opportunities and Constraints." *Renewable and Sustainable Energy Reviews* 12: 2240–2252.
Rao, N.D., and S. Pachauri. 2017. "Energy Access and Living Standards: Some Observations on Recent Trends." *Environmental Research Letters* 12, no. 2: 1–8. https://www.researchgate.net/publication/312665669_Energy_access_and_living_standards_Some_observations_on_recent_trends.
Rusdianto, E. 2014. "Keren. Geng motor keluar masuk kampung kenalkan teknologi ramah lingkungan" [So cool! Motorcycle Gangs Going In and Out of Villages to Promote Environmentally Friendly Technology]. (In Bahasa Indonesia). Mongabay. https://www.mongabay.co.id/2014/09/27/keren-geng-motor-keluarmasuk-kampung-kenalkan-teknologi-ramah-lingkungan/.
Tickell, J. 2018. *Kiss the Ground: How the Food You Eat Can Reverse Climate Change, Heal Your Body & Ultimately Save Our World*. New York: Simon and Schuster.
Tickell, J., and R. Tickell, director-producers. 2020. *Kiss the Ground*. http://www.kissthegroundmovie.com.
Wahyudi, J., and Tb.B.A. Kurnani, and J. Clancy. 2015. "Biogas Production in Dairy Farming in Indonesia: A Challenge for Sustainability." *Renewable Energy Development* 4, no. 3: 219–226.

Chapter 10
Urban Reform in Indonesia

Ahmad Rifai

Introduction

"Civic engagement" is a very broad term that takes on different meanings when seen from the perspective of different territories, localities, bodies of knowledge, and ideologies. Bringing those many perspectives into a single edited volume in a very limited time is challenging. Yet it certainly gives wider exposure to knowledge sharing and exchange. As civic engagement and participation in development is context-specific, documenting smart practices to help inform community members and encourage greater participation is preferable to formulating in the abstract what method or approach will best increase public participation.

There is an interesting notion from the civic engagement movement: we recognize the chaotic nature of liberal democracy, where things are moving beyond our control and are totally dependent on market-driven development. Urban development is moving so fast—and the community is so uprooted by the impact of development that puts such a high degree of protection on individual rights—that our social value as a community gradually disappears. Why is participation important? We want to strengthen community in order to defeat selfishness (the source of the greed of capitalism and consumerism) and build a new contract of citizenship that respects collective values, social spirit, and development that leaves no one behind.

Yayasan Kota Kita or Our City Foundation[1] is a national non-profit organization based in the city of Solo, Indonesia. The mission of Kota Kita is to support development in cities and—through the promotion of thoughtful and inclusive decision-making processes—facilitate community participation and collective action. Kota

[1] See www.kotakita.org.

A. Rifai (✉)
Kota Kita Foundation, Central Java, Indonesia
e-mail: rifai@kotakita.org

M. Indrawan et al. (eds.), *Civic Engagement in Asia*,
https://doi.org/10.1007/978-981-16-9384-7_10

Kita was founded in 2009 and currently ranks among the select urban-sector development NGOs advocating contemporary urban issues while working from grassroots to city-wide scales and engaging across the national and global scales.

The Kota Kita team has a range of different skills that relate to urban planning, architecture, urban design, community organizing, communication, and research and serves as an advocacy and resource center for urban development in Indonesia. To date, the foundation has carried out projects in 12 rapidly-urbanizing cities in Indonesia. This began with the community mapping activity in Solo (aka Surakarta), Central Java, in 2010; Kota Kita then expanded its activities in different cities. Currently, the organization works under three main scopes: urban governance and citizenship, urban inclusivity, and urban resilience.

We operate based on our key organizational values: include pedagogy, raise awareness, and encourage participation. In the different work that we do, Kota Kita collaborates with communities, citizens, city governments, national government, and international donors to bring about change and improvement in urban settings.

Beyond urban area improvement projects, Kota Kita has started work to engage with the implementation of the Sustainable Development Goals (SDGs) and the New Urban Agenda. Adopting the two into our work is exemplified by the refinement of our 'Mini Atlas,' a neighborhood-based information tool produced after a city-wide mapping process, which is now color-coded based on SDGs goals to contextualize local thinking in relation to its contributions to global agendas.

Problems and Challenges

Kota Kita believes that citizens should be at the heart of the decision-making process that determines the environment and the quality of living of the urban center they reside in. They are entitled to governance that is inclusive, transparent, and socially just. Our work over the last 10 years has aimed to meet these goals, by working to promote and strengthen participation. In doing so, we have faced the following main challenges:

- **Elite Capture:** Participatory process often only captures the elite's voices; those who are vocal in the community often dominate the process. Solutions are needed to ensure that everybody's voice is represented in the forum, particularly when the dominant voices fail to represent the whole community's needs.
- **Limited Participation of the Marginalized in Development:** Development processes often neglect the voice of the marginalized, including the poor, women, children and youth, persons with disabilities, and migrants, although their voices are *essential* in the making of an inclusive city. Some efforts to involve the marginalized in these processes are in place, but affirmative actions and policies are needed.
- **Technocratic/Bureaucratic Processes:** Lengthy processes of participation often fail to disrupt power relations between citizens and the government.

- **Limited Access to Information:** Availability and accessibility to information are essential to enable meaningful participation of citizens in development. Meanwhile, most cases in Indonesia show that basic information about the city that can help the planning process is lacking.
- **Fewer Private Sector Participants:** Much evidence demonstrates that private sector actors are not in favor of participating in public affairs.
- **Decreased Interest from Political Actors/Decision Makers:** Politicians and government officials often regard participation as a time-consuming and ineffective process. For them, participation is not worth selling in political campaigns.

What Are Kota Kita's Key Strategies for Civic Engagement? Who and What Have Been Primarily Targeted?

As mentioned above, Kota Kita collaborates with communities, citizens, city governments, national government, and international donors. Strategies for civic engagement are manifested in our three main organizational values outlined below. Specific implementation strategies are developed and then further elaborated within each project, depending on the city where it is carried out.

- **Pedagogy:** Through promotion of learning and development of tools, Kota Kita is committed to building the commitment and capacities of the next generation. Through research, capacity building, and shared pedagogic experiences from within our organization, Kota Kita continues to support actively engaged urban leaders.
- **Raising Awareness:** Kota Kita has reasoned that shared information can help address the challenges and opportunities that come with rapid urbanization. Kota Kita is therefore committed to establishing urban data and promoting the accessibility of information in order to increase awareness and promote change.
- **Encouraging Participation:** Kota Kita believes that as it relates to strengthening our communities and places, we all have roles and responsibilities for actively participating in urban development. Kota Kita facilitates and promotes the involvement of all stakeholders, especially the poor and marginalized. Only through the facilitation of different perspectives, voices, and strengths can cities be made better and decently livable.

Different types of engagement from the community level to the global scale are maintained to bring about change and improvement in the urban settings. In summary, Kota Kita is facilitating civic engagement by:

- Facilitating planning processes;
- Developing and introducing innovations to support citizen participation;
- Providing data and information to inform and empower communities;
- Facilitating dialog and acting as a bridge;
- Building capacity to advocate for change;
- Researching and producing knowledge.

Examples of Kota Kita's Civic Engagement Projects

The following are some examples from our past and ongoing projects that highlight our civic engagement experiences in carrying out participatory design processes, data-driven advocacy, facilitation, promoting learning, and developing information tools.

- **City-Wide Community Mapping, Solo, Central Java[2]:** In 2010, Kota Kita began working with then-Mayor Joko Widodo[3] and local leaders in Solo to collect data about the city's many neighborhoods. Data of importance to local development such as access to water, sanitation, poverty levels, and the number of children enrolled in school were collected from different neighborhoods and presented in mini-atlases to provide a reference for the annual participatory forum for development-planning and budgeting (*musrenbang*). This formal multi-stakeholder process supports evidence-based advocacy for improving public services and encourages data interpretation skills and self-representation. Urban planning decisions were thus transferred to the community. Facilitating the collaboration of residents in the collection and discussion of data, backed by intimate knowledge of the residents' own needs, helped governments serve the interests of their communities. In every year since 2010, Solo Kota Kita's participatory data collection and facilitation method have been undertaken to update Solo's database. Subsequently, this approach has been followed by other cities in Indonesia, including Banjarmasin, Padang, Makassar, Pekalongan, and even overseas in Ulaanbaatar, Mongolia.
- **Participatory Urban Design, Banjarmasin, South Kalimantan:** In Banjarmasin City, riverbanks are the localities for important economic and social activities. Unfortunately, rapid urbanization has given rise to chronic pollution in the water bodies and has destroyed crucial utilities in the many households. In response to the challenge, Kota Kita has facilitated a series of workshops in which residents participated and learned about water and sanitation issues, as well as the complexities of built environments, and eventually worked toward sustainable living solutions in each of their neighborhoods.

  A progressive initiative by Firm Foundation[4] facilitated participatory designs for a new waterfront public space for the neighborhood, namely a port and gateway. The structure was built over the water from a simple and resilient design and materials that community members can assemble themselves without heavy machinery. In addition to providing an enhanced public space in a neighborhood where such

[2] Solo Kota Kita city-wide mapping is a joint effort by Kota Kita, UN Habitat, USAID, Ford Foundation, and the City Government of Solo.

[3] Before Mr. Widodo proceeded to become the governor of Jakarta province in 2012, and soon after President of the Republic of Indonesia from 2014.

[4] Firm Foundation is a joint initiative by Kota Kita; AECOM UrbanSOS; and the municipality of Banjarmasin (Planning Agency; City Spatial-planning Office; PNPM Mandiri—National Poverty Alleviation Program).

spaces are rare, Firm Foundation supports informal economic activities and structural support, including a reconstructed boardwalk, a gray water vegetative filtration system, and a reestablished port for intermodal transport connections. To help connect people and public spaces, Kota Kita used storyboards, illustrations, and card games; it also employed storytelling activities and had residents tell us about the activities and uses of spaces in their daily lives. We have recorded this process and published it as a guidebook called Social Design Field Guide.

- **Urban Citizenship Academy:** The Urban Citizenship Academy (UCA) program is an initiative by Kota Kita to facilitate and support a new generation of young leaders in solving pressing urban problems. UCA serves as a platform for transmitting Kota Kita's approach and methodologies to youth in cities across Indonesia. Training consists of a series of three 2-day trainings, spaced over several months, with intensive fieldwork in between, including (1) mapping and issue identification, (2) data analysis and proposal development, and (3) development of advocacy tools and campaigns. To date, Urban Citizenship Academy has trained more than 100 youth in three cities in Indonesia and will continue the program in coming years, expanding the reach to more Indonesian cities.
- **Women on Wheels-Promoting Alternative Transportation Options for Women:** Over the past 20 years, a lack of alternative public transportation options and low-cost access to finance (e.g., easy purchase of motor vehicles on credits) have contributed to the decline of the use of bicycles in favor of motorcycles as a practical form of transport. Despite this long-term trend, over the past five years, bicycling has emerged as a recreational activity among the burgeoning middle class, whose women riders are influential and could potentially put pressure on governments for the needed infrastructure investments and pro-bicycling measures. Kota Kita designed and delivered the Women on Wheels program as a multifaceted pilot project aiming to promote bicycle use for women and girls. This was an effort to empower them and encourage them to engage equally in social and economic activities while promoting the sustainable and livable city of Solo. Through focus group discussions and case studies, we identified challenges that women face when they want to engage in cycling. We used this information to evaluate the feasibility of 'Women on Wheels' and eventually mounted the campaign aimed at encouraging women to bicycle, especially in Solo. This project started in 2015 and was completed in 2018. It incorporated various sets of actions, including educating and raising awareness about the benefits of bicycles.

The Way Forward

Strengthening and further developing three kinds of initiatives will be crucial to the success of Kota Kita's work over the coming years. First, the city-wide community mapping methodology used by Kota Kita is applicable and replicable for addressing broader urban issues. It is used as an intellectual foundation in the understanding of urban problems and designing advocacy activities. Kota Kita has applied the method to conduct data analysis, aiming to contribute to an advocacy agenda on various urban issues, such as disability access, creative economy, and promoting cycling in the city. The methodology has proven helpful to Kota Kita in encouraging more inclusive participation, thus leading to a better understanding of urban issues.

Second, the Urban Citizenship Academy (UCA), an initiative that promotes grassroots urban activism to the younger generation, has been successful so far due to a widespread rise in the interest of youth in volunteerism activities. However, we realize that improvement is still needed for a more impactful result. The training process of UCA requires mechanisms for mentorship. However, we might be lacking in mechanisms for assistance and mentoring. Lessons learned from this process include the need for improvement in the methodology and assistance for further practice of the UCA process. What is needed is to better connect the group of participants to the resources required in order for them to execute the proposed projects, and to give more intensive assistance for producing a better quality advocacy strategy.

Third, the Right to the City concept, which recognizes cities as a common good, advocates a strong message about the collective right of citizens to urban areas. The Right to the City (R2C) movement, since its engagement in PREPCOM III Habitat III in Surabaya in 2016 (Habitat III 2016a), has played important roles in promoting the concept in Indonesian cities. With the support of the Global Platform Right to the City (GPR2C 2018), Kota Kita wants to further strengthen the perspectives of the right to the city in regional Southeast Asian countries. The adoption of the New Urban Agenda (UN SDGs 2016; Habitat III 2016b) will bring more opportunities for discussing "inclusive city" and "no one left behind."

Visions and Plans for Regional Collaboration

Representing civil society organizations, Kota Kita has been proactively involved in the Habitat III advocacy in Indonesia since our activities began in Surabaya, 2016. Our current advocacy is focused on increasing support for the Right to the City in the New Urban Agenda and (re-)introducing the concept to Indonesian and Asian networks as a means to reform the city and civic rights in our region. Our annual Urban Social Forum (USF)[5] also acts as a platform for many urban and social activists to meet, interact, and develop collaborations, realizing 'Another City is Possible!' The event in 2018 recorded as many as 1,500 participants from Indonesian and Asian cities. Kota Kita anticipates that we will be able to promote the discourse of sustainable urban development in Indonesia and Asia, through strengthening the civil society movement in the region.

[5] The yearly one-day gathering toward improving cities, led by civil society. This Indonesian forum rallies NGOs, community-based organizations, practitioners, students, and social leaders. The slogan of the gathering, "Another City is Possible!", reflects the aim of challenging urban dwellers to re-imagine the city by exploring alternative policies and initiatives to mitigate development challenges in the city.

References

GPR2C. 2018. "Global Platform for the Right to the City: What Is It?" GPR2C Website. http://www.righttothecityplatform.org.br/o-evento/.

Habitat III. 2016a. "Preparatory Committee III (PrepCom3)." Habitat III Website. http://habitat3.org/the-new-urban-agenda/preparatory-process/preparatory-committee/prepcom3.

Habitat III. 2016b. "The New Urban Agenda." Habitat III Website. http://habitat3.org/the-new-urban-agenda.

UN SDGs. 2016. "The New Urban Agenda: Key Commitments." UN SDGs Website. https://www.un.org/sustainabledevelopment/blog/2016/10/newurbanagenda/.

Chapter 11
A Decade of Fighting Box Jellyfish Health Issues

Lakkana Thaikruea

Highlights

- When unknown and complex health problems strike, experts need to be prepared to step out of their "comfort zone" and systematically engage with society in order to secure long-lasting solutions that combine research and outreach.
- Science and local knowledge can be brought together by virtue of longer-term commitment.

Introduction

The previously little-known box jellyfish (the class *Cubozoa*) has become a menace to the coasts of Thailand. The venomous stinging jellyfish had been investigated at local and national levels since 2008 and was verified in 2009 by a medical epidemiologist from the Ministry of Public Health of the Thai government and the author (the two of whom comprised the leading team). It gradually became known to the tourism industry and health stakeholders. The lethal sting of box jellyfish can kill humans in 2–10 min (Fig. 11.1). Prior to the work of the leading team, uninitiated health personnel misperceived and misdiagnosed cases, and experts provided information to government officers that Thailand did not have such lethal jellyfish and even denied

The author's initiatives with regard to lethal jellyfish have been recognized by multiple awards (Thailand Outstanding Preventive Medicine Expert of Epidemiology-2019; Surveillance and Prevention from MNRE-2015; Best Case Study from Engagement Thailand-2015; and Box Jellyfish Expert from MOPH and MNRE-2014)

L. Thaikruea (✉)
Department of Community Medicine, Faculty of Medicine,
Chiang Mai University, Chiang Mai, Thailand
e-mail: lakkana.t@cmu.ac.th

M. Indrawan et al. (eds.), *Civic Engagement in Asia*,
https://doi.org/10.1007/978-981-16-9384-7_11

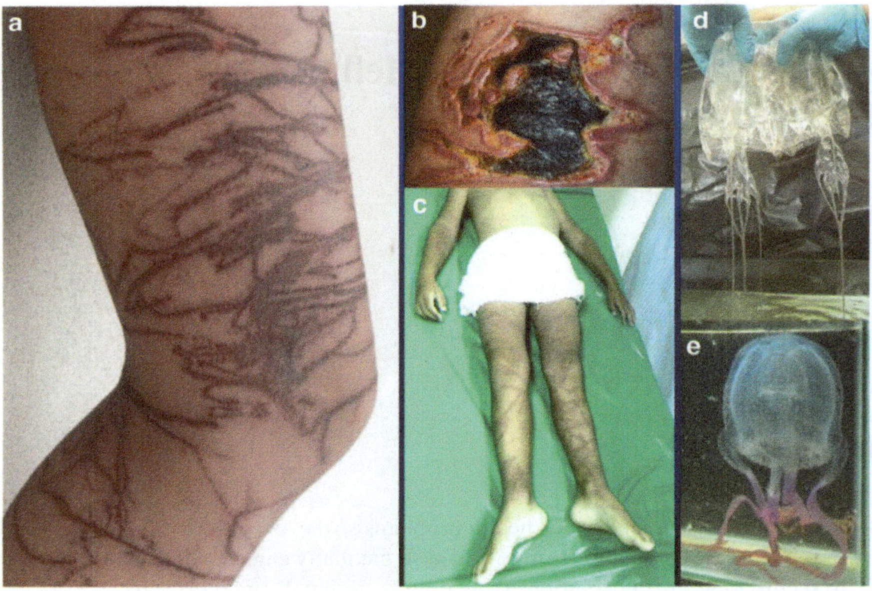

Fig. 11.1 **a**: Caterpillar track like of tentacle marks on the leg, **b**: Necrosis wound on the thigh, **c**: Fatal case, **d**: Multiple tentacle box jellyfish, and **e**: Single tentacle box jellyfish

the presence of toxic jellyfish. Misunderstanding and the lack of data and knowledge led to controversies, and the international media claimed there had been a cover-up of the problem. With lasting solutions that combine persistent research and outreach, ways were found to control the risk by determining the proper first aid applications and appropriate prevention measures. Ensuing activities led by the leading team continued with establishment of a task force, joined by a team of dedicated experts and communities. The successful establishment of toxic jellyfish surveillance and networks, as well as continued public health education and risk communication followed. Due to these efforts, there has not been a lethal case since 2016. This is a success story that managed to turn a controversial health threat with notable conflicts of interest among stakeholders into national and international solutions.

The Rise of the Problem

In 2008, a medical epidemiologist from the Epidemiology Division of the Ministry of Public Health of the Thai government received a notification of an unexplained death of a tourist. She invited the author, as a medical epidemiologist from the Faculty of Medicine, Chiang Mai University (Med-CMU), to form a leading team. The leading team investigated and found that the death might have been caused by a box jellyfish sting. The small team later developed a task force dedicated to undertaking research and outreach of the previously little known toxic jellyfish that invaded the coast of Thailand.

At the time, the situation was difficult and complicated because of conflicts of interest, misdiagnosis (most of the diagnoses suggested an anaphylactic shock or in layman's terms severe allergy to jellyfish) that led to inappropriate treatments and underreporting, the lack of effective first aid, inappropriate traditional treatments (such as rubbing the wound with mashed morning glory leaves mixed with water or alcohol, rubbing the wound with sand, and pouring rain water or alcohol on the wound) that led to death, and the lack of laboratories and expertise (Thaikruea et al., 2012).

At least eight fatal cases were found in Thailand between 1999 and 2015. Four envenomation victims died within two minutes, three died within one to ten minutes, and one within a few minutes (Thaikruea et al. 2015; Thaikruea and Siriariyaporn 2016, 2018, chaps. 1, 2, and 3). Envenomation means a physical discharge of venom that affects the body into the victim's bodily tissues. The venom is comprised of a complex mixture of polypeptides and proteins, including hemolytic, neurotoxic, cardiotoxic, and dermatonecrotic toxins (Lau et al. 2019; Fenner and Harrison 2000; Thaikruea and Siriariyaporn 2015; Leelarasamee and Thaikruea 2018; Thaikruea and Santidherakul 2018; Thaikruea and Siriariyaporn 2018, chaps. 3 and 4; Tibballs 2006).

Box jellyfish are comprised of two orders, namely *Chirodropida* and *Carybdeida*, and eight families. They are among the most lethal in the world. One species, *Chironex fleckeri*, is particularly dangerous. Its envenomation may cause the victim to stop breathing and his/her heart to beat with "rapid cardiorespiratory depression," possibly leading to death in less than 10 min (Lau et al. 2019; Fenner and Harrison 2000; Leelarasamee and Thaikruea 2018; Tibballs 2006).

The toxic jellyfish outbreak compromised the safety of human populations on the coastal areas of Thailand, which represented 14 million of the country's 63 million residents in 2008.[1] The areas were also visited by numerous national (226 million) and international (76 million) visitors, in 2018.[2] Conflicts of interest and denial of the existence of lethal jellyfish ensued.[3]

---

[1] http://stat.bora.dopa.go.th/stat/y_stat51.html [Accessed March 16, 2019].

[2] https://thai.tourismthailand.org Thailand was visited by 14,584,220 and 34,431,489 foreign tourists in 2008 and 2018, respectively. http://statbbi.nso.go.th/staticreport/page/sector/th/17.aspx [Accessed March 19, 2019] and https://www.mots.go.th/more_news.php?cid=411 [Accessed March 19, 2019].

[3] Conflicts of interest included: (1) Hotels/resorts did not allow installation of warning signs and vinegar first aid poles because they feared that they would scare tourists and make them not want to stay; (2) Local community members, in particular those working in the tourism industry, did not want the mass media to publicly report on incidents because they were concerned that doing so would reduce the number of tourists; (3) Tourists wanted to know so that they could adjust their risks and prepare themselves; (4) Mass media wanted to report the incidents and the causes of near-fatal and fatal cases; (5) The foreign embassies in Bangkok wanted to know the cause and magnitude of the problem in order to warn their nationals; and (6) the Ministry of Public Health of the Thai government wanted to know the cause in order to develop control and prevention measures.

Contributing to the denial of the existence of lethal jellyfish, marine biologists and health professionals initially believed that this deadly box jellyfish species existed in other countries (such as *Chironex fleckeri* in Australia) but not in Thailand. Furthermore, no one had adequate information about lethal box jellyfish and no expert on box jellyfish envenomation could be found in Thailand in 2008. A laboratory to identify species and toxins has become available only in recent years, and only for research purposes. Its capacity is not sufficient for public service. Between 2008 and 2011, the local communities and other stakeholders were reluctant to even acknowledge that deaths caused by jellyfish occurred, and because of the lack of knowledge as well as surveillance systems, no effective intervention was made (Thaikruea 2016, 26–29; 2018; Thaikruea and Siriariyaporn 2018, chap. 6).

Initial confusion was multifaceted. For instance, in keeping the beach goers informed, warning boards were constructed, which said "bluebottles," when they should have said "box jellyfish." The signboards did not even present an illustration of box jellyfish, nor instructions for how to avoid or treat a sting. At one stage, a marine biologist announced through the media that the proper first aid would be to splash seawater on the wound before applying vinegar. The seawater, however, is not necessary. In fact, a force of splashing can trigger nematocyst firing (stinging cells contain microscopic organelles that discharge explosively, injecting a toxin) and so may exacerbate the problem. The limited capacities, in terms of financial and human resources, knowledge for prevention and control, and limited education materials continued up to 2014 (Thaikruea 2018; Thaikruea and Siriariyaporn 2018, chap. 6). The issue soon became politically sensitive. Increased attention,[4] especially in the form of cavalier news coverage by the international media, stigmatized the image of the safety of beach tourism in Thailand (Thaikruea et al. 2015; Thaikruea and Siriariyaporn 2015, 2016, 2018, chaps. 2 and 6).

Through all the controversies, the task force pressed on and eventually discovered that whereas inappropriate treatments could easily lead to death, quick application of first aid could increase the probability of survival (Leelarasamee and Thaikruea 2018; Thaikruea and Siriariyaporn 2018, esp. chaps. 2 and 4). Proper application of first aid, which is carried out at the scene, is crucial for surviving the sting. By the year 2012, the task force launched the installation of vinegar stations, which look like a red pole, for first aid. Food grade vinegar (4–6% acetic acid) was found to be effective. It prevented nematocysts from explosively discharging and injecting toxins (the nematocysts still contain toxin but cannot discharge) (Thaikruea and Siriariyaporn 2018, chaps. 4 and 6); Thaikruea and Santidherakul 2018).

Furthermore, if health personnel were open to accepting the medical realities of envenomation by box jellyfish and were trained in proper treatment, jellyfish stings could immediately be treated at any health facility. Therefore, the leading team provided training and educational materials based on their research findings and experiences.

---

[4] For example, an Australian boy was stung by box jellyfish and almost died. One of his relatives posted stories about the boy on the TripAdvisor website to warn other tourists https://www.tripadvisor.com.sg/Profile/DCB95 [Accessed December 5, 2018].

Responses and Initiatives

The leading team developed a response with four key strategies: (1) Implementing research combined with outreach to determine effective intervention protocols; (2) Devising and implementing early warning and rapid response; (3) Building a strategy into the system; and (4) Developing longer-term sustainability into the response.

Convincing key stakeholders that the *Chironex* spp. of box jellyfish contains a deadly toxin and occurs along the coast of Thailand was an important first step. The leading team provided evidence about fatal cases via different channels for each target audience, such as video clips and mass media news stories for the communities and the other stakeholders, scientific articles published in international journals for experts and health professionals, and official reports of fatal case investigations for policymakers. The author found that although many stakeholders in high risk areas of Samui and Pha-ngan islands believed that box jellyfish could injure or kill humans, some people remained skeptical or thought the problem was insignificant. The leading team obtained and provided hard evidence to convince stakeholders to implement sustainable prevention measures. The ensuing retrospective study of severe cases involving admission into hospitals or health clinics in Samui and Pha-ngan islands, covering 1997–2015, showed at least 15 severe box jellyfish cases, especially during August–October. In reviewing all these recorded incidents and the wounds that developed immediately after the sting, it was found that in more than half of the cases the victims were rendered unconscious. In six of these incidents, fatalities occurred (46.7%) (Thaikruea and Siriariyaporn 2016).

The leading team conducted a diverse range of research. The team used epidemiological knowledge, including outbreak investigations, research, surveillance, evaluation, and monitoring, to gather evidence to convince the relevant stakeholders and to form a base for the development of strategic plans. Field assessments of the magnitude of the problem and affected populations were also needed to develop preventive measures in the long run.

The leading team considered community responses to box jellyfish stings to be an important part of the mitigation strategy. Rapid response from community members is critical given that the "golden period" before the envenomation led to death ranged from only 2 to 10 min (Thaikruea and Siriariyaporn 2015, 2018, chap. 4). The stratagem, therefore, combined scientific investigation with communities' own experiences, along with local wisdom in order to build the necessary evidence base for effective response, including communication of risks. The strategy also called for inclusive constituency building, including cooperation between scientists, communities, and government at multiple levels, as well as the effective mobilization of resources.

The surveillance and outbreak investigation included development of guidelines (prevention, first aid, and treatment)[5] and therefore increased the scientific

[5] The first aid includes: (1) Pour vinegar with 4 to 6% acetic acid immediately and continuously throughout the wound for at least 30 s and do not pour fresh water or alcohol; (2) Call for help or call ambulance; (3) Someone should stay with the injured because he/she may be unconscious

and applied knowledge necessary for dealing with box jellyfish stings. Eventually communities' capacities also increased, through dedicated educational materials, covering jellyfish, jellyfish envenomation, signs and symptoms, diagnosis, nematocyst, first aid, treatments, complications, surveillance, and prevention, and control, alongside training course materials which covered the same topics (Thaikruea 2018; Thaikruea and Siriariyaporn 2018; Leelarasamee and Thaikruea 2018).

The early warning and rapid response took considerable time and effort to develop. In the beginning, the team had to establish surveillance in an ad hoc manner to detect cases. Early warning systems developed slowly and gradually, and now involve a combination of approaches including the following:

1. Active: Authorized staff actively find a case for early detection (costly and time-consuming);
2. Passive: The information is routinely sent to the authorized staff (slow and underreported);
3. Event-based: These approaches increase the sensitivity of responses by rapidly capturing information about events that are a potential risk to public health, e.g., through word of mouth, news events related to the occurrence of symptoms, incidences of sudden death during marine activities, and so on;
4. Indicator-based: This is the systematic collection, monitoring and interpretation of structured data (usually passive). e.g., medical reports of unusual disease patterns;
5. Community-based: An active process of community participation in detecting, reporting, responding to and monitoring health events in the community, focusing on systematic on-going collection of data on events and diseases. Case definitions and forms are simple for reporting to health facilities. The authorized staff verify, investigate, collect, analyze, and respond.

For surveillance of injuries and deaths caused by toxic jellyfish, however, the leading team had to take the following factors into consideration: (1) local cultural and social contexts, including various beliefs and dynamics within affected communities; (2) risks such as rapid death; (3) inappropriate traditional first aid treatments; (4) limited financial and human resources; and (5) the general lack of knowledge regarding jellyfish envenomation. In the beginning, the leading team established ad hoc surveillance in order to detect suspected cases and build knowledge. The team improved the ad hoc surveillance by increasing case detection (case definition and outbreak investigation) and establishing three toxic jellyfish networks, namely a task force, experts, and communities (Fig. 11.2). This ad hoc surveillance system was then incorporated into the national surveillance system. For community-based surveillance, the team encouraged a sense of responsibility and ownership, supported processes that increase coordination and cooperation, and used locally available resources to support timely notification and action. To avoid false warnings about

within a few minutes; (4) Do not rub the wound; and (5) If the victim becomes unconscious, does not breathe, or has no pulse start cardio pulmonary resuscitation first. As for preventive measures, tourists are advised to swim inside a stinger net, if readily available, wear a tight full body swimming suit and avoid swimming during the night time.

Fig. 11.2 Toxic Jellyfish networks and stakeholders

toxic jellyfish sightings and stings, mainly from event-based and community-based surveillance, the leading team and marine biology experts served as core members of all networks to quickly verify purported incidences involving toxic jellyfish from the starting step of notification (not passively waiting for reports to be analyzed later). The task force also collaborated with marine biologists and ecologists mainly from the Marine and Coastal Resources Division of the Ministry of Natural Resources and the Environment (MNRE) and international, mostly Australian, health experts. MNRE supported a budget for installing vinegar poles and educational warning signs. False positives were investigated by verifying whether the jellyfish that networks reported (from beaches or sea) were health threatening or toxic.

The leading team has educated members, in particular fisherfolk, tour guides, long-tail boat drivers, divers, hotel staff, jet ski riders, resort owners, health and non-health personnel, and the tourist police, via toxic jellyfish networks and by providing educational materials in addition to other formats such as meetings, training, workshops, supervisions (concerning outbreak investigation, diagnosis, treatment, prevention, and control programs), and websites (Figs. 11.3 and 11.4). The urgent intervention activities were also done via the networks. The ultimate goal of surveillance was to take timely action. Prevention and control measures were done not only at local but also national and international levels. The team leveraged community resources and structures for early warning and rapid response. Prompt national-level assistance was necessary in order to protect the community's health (Fig. 11.5).

Fig. 11.3 Training various groups and vinegar pouring

Fig. 11.4 **a**: A vinegar educational warning sign and second model stinger net, **b**: A second model of stinger net maintenance, **c**: Multiple box jellyfish were caught outside the stinger net, **d**: A second model of stinger net monitor, **e**: A vinegar first aid pole

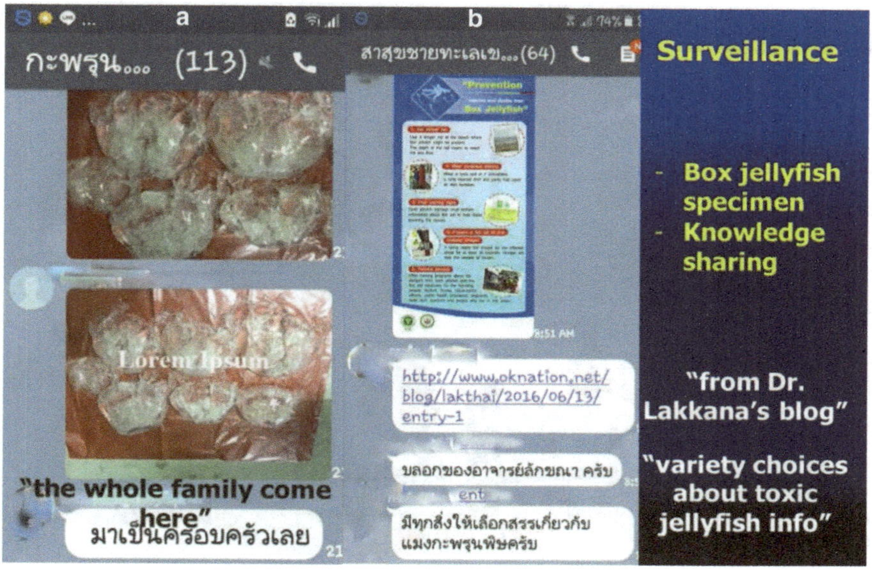

Fig. 11.5 Example of toxic jellyfish surveillance in a community on Pha-ngan island. **a**: The box jellyfish were collected outside stinger nets during daily monitoring. The resort owners reported the event and the local health workers sent the specimen to the experts, **b**: The network also shared knowledge via social media, i.e., Line

In summary, the surveillance system of injuries and deaths caused by toxic jellyfish has a combination of active, passive, event-based, indicator-based, and community-based elements. The resulting early warning system (EWS) showed promising effectiveness. Traditional surveillance was dependent mainly on indicator-based approaches, and therefore the current system of combined approaches increased sensitivity and accuracy. The EWS was also progressive in that it allows strong multi-stakeholder participation. The bottom up approach combined with the top down role of authorization and expertise helped strengthen the capabilities for early warning by reducing cases of false warnings and provided risk communications as well as rapid responses. The system also fostered multidisciplinary approaches and was therefore scientifically more robust. After the EWS was established, occurrences of deadly jellyfish were known, and detection improved. The quality of reports of injuries and fatalities caused by the jellyfish increased, due to improved early warning and rapid response. By 2016, there appeared to be no fatalities.

In building the response system, evidence-based management was used to inform all decision-making, plans, and execution. Sources of evidence included the findings from field research, as well as toxic jellyfish surveillance systems and outbreak investigations. The data were collected, analyzed, interpreted, translated, and disseminated to the target audience. The target audience included three groups: policymakers, community/stakeholders, and the general population (Fig. 11.6).

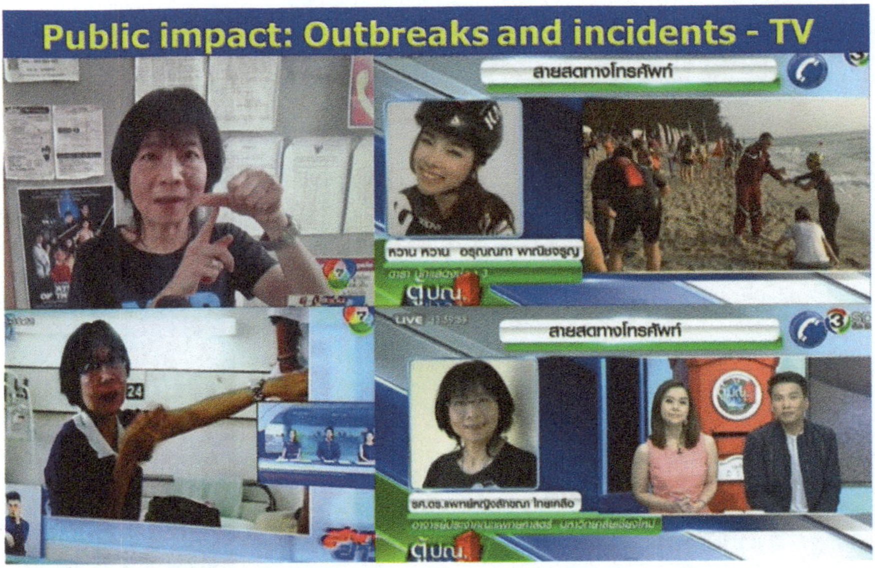

Fig. 11.6 Public impact: The author provided knowledge and risk communication to the mass media when there were outbreaks, incidence cases, or national and international triathlon events involving injuries from toxic jellyfish

Sustainability was developed on the principles of egalitarian partnership and policy support by the Thai government at the national and local levels (Thaikruea 2016, 26–29; 2017, 87–120; 2018; Thaikruea and Siriariyaporn 2018, chap. 6). One example of this can be seen in the collaboration by the Ministry of Public Health and Ministry of Natural Resources and the Environment in developing MoUs addressing research into and response to the health risk constituted by the jellyfish (further details below).

Stages of the Responses

The chronology of the responses consisted of three stages (Thaikruea 2016, 26–29; 2017, 87–120; 2018). The first of these stages involved development of the information base and surveillance infrastructure between 2008 and 2011. The main initial mission was to prove the existence of deadly jellyfish and to assess the magnitude of the problem.

Case investigation and research were conducted to learn basics such as clinical manifestations, first aid, treatment, prevention, and control. Ad hoc surveillance systems and two toxic jellyfish networks (task force and expert team networks) were established in 2008 to serve these activities. The initial members of the task force and expert team came from national and international circles including Australian universities and the Divers Alert Network (DAN), as well as an Australian journalist. The third toxic jellyfish network of communities was soon established and expanded to include other relevant stakeholders such as resort/hotel managers and owners, speed boat/long-tail boat operators, divers, and biologists. Through interviews and discussions, these primary stakeholders contributed to the collection of detailed information, as well as building the bases of collaboration for prevention. Toxic jellyfish field surveillance of venomous jellyfish incidents began in 2009 (Thaikruea 2009). The task force prospectively and retrospectively investigated all suspected cases of box jellyfish sting outbreaks, which were detected by the toxic surveillance system and network.

The second stage, from 2012 to 2014, involved gathering knowledge for improving medical care (diagnosis and treatment) and responding to public health risks (prevention and control measures), systems improvement, and integration of ad hoc surveillance into the national surveillance system. This system was known as the "Injuries and Deaths Caused by Toxic Jellyfish - Surveillance System." MOPH and MNRE included prevention and control measures into fiscal year plans and expanded to other risk areas. The author assisted and provided recommendations to develop the first memorandum of understanding (MoU) between MOPH and MNRE. The good outcomes of the first MoU led to the second MoU for the time period from 2014 to 2018. The third MoU, which covers a four-year period, has more collaboration from nine related organizations and was signed in 2019 (Thaikruea 2019). The collaboration covers surveillance systems, prevention and control measures, risk communication, research, and management.

The steering team expanded the networks concerning toxic jellyfish and further facilitated research and outreach, as well as careful field implementation. The task force continued the monitoring and evaluation of toxic jellyfish incidences. The findings contributed to improving outbreak investigations, especially case definition and reporting, and outbreak investigation. The members of each network recruited more members from risk areas and from relevant organizations, including academics, tourism associations, the mass media, local governments, guide associations, human resource organizations, and long-tail boat associations. The task force identified sensitive issues (e.g., patient privacy), released locations of box jellyfish surveys to the media affecting resorts and hotels, provided press releases about the misidentification of toxic jellyfish species and ineffective first aid practices, and continued to provide important information, consultancy, and recommendations for the benefit of the network and all the stakeholders (Thaikruea 2015; Thaikruea and Siriariyaporn 2018). They improved educational materials based on field experiences, evaluation, and research findings. Through continuous knowledge management, the local communities were encouraged to develop and install vinegar first aid stations and educational signs using available resources. The first model of a stinger net to prevent jellyfish during recreation activities was designed for small business resorts. It was cheap but not durable. Examples of available resources were polyvinyl chloride (PVC) pipe, wooden board, handmade fishnet, homemade cement-filled tire, handmade cement-filled PVC pipe, etc. The knowledge was shared with other communities through networks, training, or meetings.

Based on knowledge gained in the past decade, the author incorporated improved diagnosis, early warning, and rapid response into her design of a new but simple—i.e., practical, cheap, and durable—technique named "Vacuum sticky tape technique to identify a toxic jellyfish class" (Thaikruea and Santidherakul 2018). The technique is used for collection and transfer of jellyfish specimens and eventual nematocyst identification (Fig. 11.7). The leading team also published a new reference electronic book, according to new knowledge gained, researchers' findings, and experiences, called *Injuries and Deaths Caused by Toxic Jellyfish: Surveillance, Prevention, and Treatment* in 2018 (Thaikruea and Siriariyaporn 2018; Thaikruea 2019). Its contents cover all aspects for various readers (Thaikruea and Siriariyaporn 2018).

The third stage, between 2015 and 2019, mainly concerned development of sustainability and filling in knowledge gaps. In order to be sustainable, considerable ground had to be covered through development and institutionalization of surveillance, formal government support, capacity building, resources mobilization, and partnership building (Thaikruea 2017, 87–120; 2018; Thaikruea and Siriariyaporn 2018; Thaikruea 2019).

By now, the ad hoc toxic surveillance system has improved and been permanently integrated into the national surveillance system. The necessary support by the government was further realized through established formal collaborations, including between MOPH and MNRE.

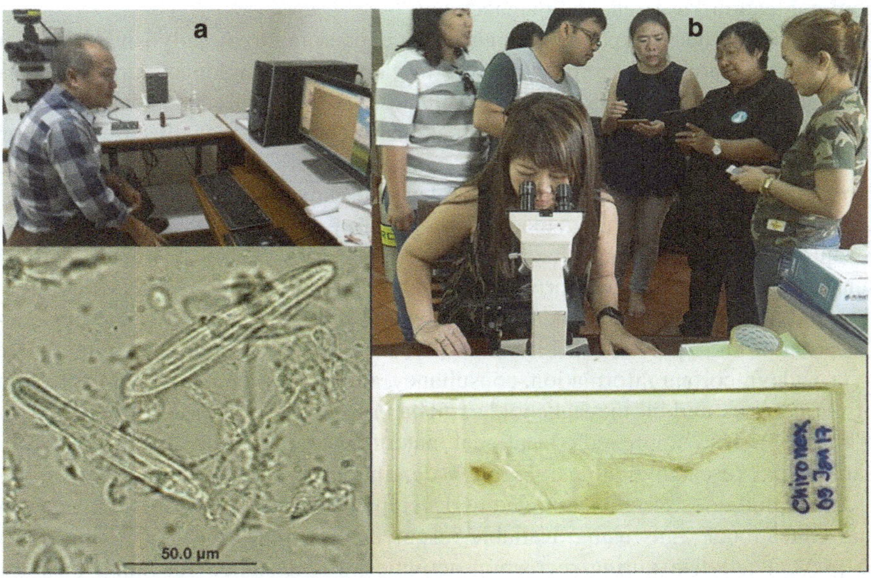

Fig. 11.7 **a**: Nematocyst identification conducted in Faculty of Medicine, CMU, **b**: The technique can be taught in resorts/hotels using local facilities

In building capacity, the task force regularly engaged with multiple stakeholder groups, including health and tourism personnel and the local communities, in order to create, evaluate, and update educational materials that are appropriate for various audiences, along with new knowledge gained through field and research experiences and training of trainers for target groups. By this time, financial resources were diversified, and included the government fiscal budget, as well as contributions from NGOs, the private sector, the local communities, and the other stakeholders.

Building partnerships has remained among the top priorities. Since the communities and the other stakeholders played an important role, the task force made concerted efforts to maintain good relationships with them and continued to engage with and support them. The intervention is now designed to fit into the local environment and utilizes available affordable resources in the communities, including vinegar first aid poles, vinegar educational signs, educational warning signs containing positive messages, and stinger nets. This is in sharp contrast with typical traditional warning signs with alarming illustrations yet without any instructions or advice to apply first aid (Thaikruea and Siriariyaporn 2018, chap. 6).

The toxic jellyfish networks continue to expand and members continue to share knowledge and experiences (Thaikruea 2016, 26–29; 2017, 87–120; 2018; Thaikruea and Siriariyaporn 2018, chap. 6). The initiative is now being replicated and disseminated not only by the government sector, but also by multiple stakeholders, including the private sector and NGOs. Toxic jellyfish networks at national and international levels were also built up. The vinegar first aid poles, which were originally developed on the small island of Mak, have now been replicated and disseminated in many islands and provinces along the coasts of Thailand (Thaikruea 2012a, b, Thaikruea and Siriariyaporn 2018). Anyone can access and transfer educational materials and all relevant information, which are available through the author's blogs, websites of partners and stakeholders, and the toxic jellyfish networks' social media platforms (Thaikruea 2016, 26–29; 2017, 87–120; 2018; Thaikruea and Siriariyaporn 2018, chap. 6).

At the current stage, the educational, communications, and technical aspects (such as the use of stinger nets) are maturing. Educational materials continue to be disseminated via social media, electronic mail, websites, YouTube, and blogs.[6] A series of trainings to update knowledge and trainings for trainers were conducted to further distribute knowledge, maintain good practices, build up local knowledge bases, and improve surveillance in the long run. The stinger net models were examined and improved. The first model was designed for small business groups, and the second model was for moderate to large business groups. The second model used durable and stronger materials that make it more effective than the first model. Thus, it costs more money (but cheaper than that of the prototype). These models were evaluated and improved each year.

[6] For example, this blog for the general public: http://oknation.nationtv.tv/blog/peeguay.

Conclusions

Since incidences of jellyfish-induced deaths ceased by 2015, increasing evidence suggests that persistent responses to a complicated lethal health threat with conflicting stakeholder interests have been successful. Under normal circumstances, public health services were provided by the Ministry of Public Health. In the toxic jellyfish case, however, effective cross-sectoral cooperation, especially with the Ministry of Higher Education, Science, Research and Innovation, MOPH, and MNRE, developed over time. Furthermore, the governing authorities have created space for wider stakeholder groups, including the local communities. Personnel originally not trained in medicine, along with trained health workers, are now able to conduct first aid and educate the general public. This is particularly important in situations where emergency personnel and/or ambulances cannot quickly reach the scene of incidents and no lifeguard is on duty at a given beach.

The impacts of this multi-pronged endeavor for dealing with toxic jellyfish were therefore multifaceted. Firstly, in terms of medical science, there is now better knowledge and clinical practice for dealing with jellyfish envenomation. This includes wound complications, diagnosis and treatment, and appropriate first aid with vinegar only. Additionally, public health aspects of venomous jellyfish stings increased in terms of improved knowledge of outbreak investigation, case detection, early warning, rapid response, as well as the establishment of comprehensive and progressive surveillance systems. Prevention and control measures included concrete recommendations concerning safety protocols for swimming at sea. Furthermore, this endeavor proved how participation by multiple stakeholders can be effective on the ground. Better biological knowledge now exists regarding lethal jellyfish, including nematocyst identification. The box jellyfish species found in Thailand are different from those in other countries. Along the way, a new species of fatal box jellyfish "*Chironex indrasaksajiae* Sucharitakul" was discovered in 2017 by a CMU student, who was one of the author's former patients (Sucharitakul et al. 2017). He was stung by box jellyfish and the author subsequently encouraged him to study the species (Thaikruea and Siriariyaporn 2018, chap. 1). The toxin of this species has been in the identification process since 2018.

The benchmark is now set. The research, surveillance, and outreach which involved knowledge building resulted in crucial outputs such as educational materials and training modules that helped risk communication, prevention, and control. Along the way, university engagement increased, thus adding to capacity building for epidemic responses. In addition, the tourism industry benefitted from the risk communications and knowledge about how to deliver first aid in the event of box jellyfish stings.

References

Fenner, P.J., and S.L. Harrison. 2000. "Irukandji and Chironex Fleckeri Jellyfish Envenomation in Tropical Australia." *Wilderness & Environmental Medicine* 11, no. 4: 233–240.

Lau, M.T., J. Manion, J.B. Littleboy, L. Oyston, T.M. Khuong, Q.P. Wang, D.T. Nguyen, D. Hesselson, J.E. Seymour, and G.G. Neely. 2019. "Molecular Dissection of Box Jellyfish Venom Cytotoxicity Highlights an Effective Venom Antidote." *Nature Communications* 10, no. 1: 1655.

Leelarasamee, A., and L. Thaikruea. 2018. "Which First Aid Treatment Is Appropriate for a Bizarre Skin Lesion and Cardiovascular Collapse after Swimming in the Sea?" *Journal of Medical Association of Thailand* 101, no. 8: 1143–1144.

Sucharitakul, P., S. Chomdej, T. Achalawitkun, and I. Arsiranant. 2017. "Description of Chironex indrasaksajiae Sucharitakul sp. nov. (Cnidaria, Cubozoa, Chirodropida): A New Species of Box Jellyfish from the Gulf of Thailand." *Phuket Mar Biol Cent Res Bull* 74: 33–44.

Thaikruea, L. 2009. "Warning for Box Jellyfish." The Epidemiologist Life Blog website. http://oknation.nationtv.tv/blog/peeguay/2009/02/17/entry-1.

Thaikruea, L. 2012a. "Amazing Koh Mak in Dealing with Toxic Jellyfish Health Threat—Part 1." The Epidemiologist Life Blog website. http://oknation.nationtv.tv/blog/peeguay/2012/05/21/entry-1.

Thaikruea, L. 2012b. "Amazing Koh Mak in Dealing with Toxic Jellyfish Health Threat—Part 2." The Epidemiologist Life Blog website. http://oknation.nationtv.tv/blog/peeguay/2012/05/21/entry-1.

Thaikruea, L. 2015. "Breaking News! Probable Injured and Fatal Cases Stung by Box Jellyfish." The Epidemiologist Life Blog website. http://oknation.nationtv.tv/blog/peeguay//2015/08/01/entry-1.

Thaikruea, L. 2016. *University Engagement: A Case Study of Thailand's First Toxic Jellyfish Networks and Knowledge Establishment*. Bangkok: Civil Society Engagement Thailand.

Thaikruea, L. 2017. *Strategy of Project with Multidisciplinary and Multiorganization and Community Involvement: How to Use Simple Techniques for Developing a Strategic Plan*. Chiang Mai: Faculty of Medicine, Chiang Mai University Publishing.

Thaikruea, L. 2018. "A Decade of Fighting the Lethal Jellyfish Health Problem." Presentation at the 4th Asiaengage Regional Conference, the Impress Hotel. Chiang Mai, Thailand, November 26–28, 2018.

Thaikruea, L. 2019. "Strategies and Success in Fighting the Lethal Jellyfish Health Threat in Thailand: 2008–2019." In Workshop session of Social Engagement as a Tool for Disruption—Lessons from Thailand at NCCPE Conference-Engage 2019: Disrup. 2019, the Bristol Marriott Hotel City Centre, Bristol, United Kingdom, December 4–5, 2019.

Thaikruea, L., and S. Santidherakul. 2018. "The Public Health Impact of a New Simple Practical Technique for Collection and Transfer of Toxic Jellyfish Specimens and for Nematocyst Identification." *Journal of Public Health Policy* 39, no. 2: 143–155.

Thaikruea, L., and P. Siriarayaporn. 2015. "Severe Dermatonecrotic Toxin and Wound Complications Associated with Box Jellyfish Stings 2008-2013." *Journal of Wound, Ostomy & Continence Nursing* 42, no. 6: 599–604.

Thaikruea, L., and P. Siriarayaporn. 2016. "The Magnitude of Severe Box Jellyfish Cases on Koh Samui and Koh Pha-ngan in the Gulf of Thailand." *BMC Research Notes* 9: 108.

Thaikruea, L., and P. Siriariyaporn. 2018. *Injuries and Deaths Caused by Toxic Jellyfish: Surveillance, Prevention, and Treatment*. Chiang Mai: Faculty of Medicine of Chiang Mai University. D-Library/National Library of Thailand. http://164.115.27.97/digital/items/show/12987.

Thaikruea, L., P. Siriariyaporn, R. Wutthanarungsanm, and P. Smithsuwan. 2012. "Toxic Jellyfish Situation in Thailand." *Chiang Mai Medical Journal* 51, no. 4: 93–102.

Thaikruea, L., P. Siriariyaporn, R. Wutthanarungsan, and P. Smithsuwan. 2015. "Review of Fatal and Severe Cases of Box Jellyfish Envenomation in Thailand." *Asia-Pacific Journal of Public Health* 27, no. 2: 1639–1651.

Tibballs, J. 2006. "Australian Venomous Jellyfish, Envenomation Syndromes, Toxins and Therapy." 48, no. 7: 830–859.

Chapter 12
Citizens' Initiatives in the Fukushima Radiation Disaster: Measuring and Sharing Fukushima

Mariko Komatsu

Triggered by a massive earthquake and tsunami on March 11, 2011, the Fukushima Daiichi Nuclear Power Plant accident released a large volume of radioactive material and spread invisible radiation via air circulation, rain and snowfall in a wide area of eastern Japan. People were suddenly forced to make changes and critical life decisions–including whether they should stay or evacuate or should continue farming and consuming their local agricultural products or abandon these altogether.

The key information residents needed to know was the level of contamination in their surroundings and foodstuffs. However, the government was unable to provide all the required information at once. Measuring radiation requires special devices that cater to specific purposes. Portable measuring devices, such as a Geiger counter, a survey meter, or a NaI scintillation detector can be used to measure the ionized radiation on a contaminated surface or in the air. There also is a badge or necklace-type counter that is worn by the residents. These devices keep track of an individual's daily radiation exposure level. For measuring the radiation contained in food items, one needs to install a measuring device weighing over 100 kg, along with a computer for analysis. Each machine requires particular maneuvers, techniques, and knowledge about radiation in order to properly install, operate, and interpret the data. These preparations and equipment were not readily available because a radiation disaster like this with such impacts was "unforeseen," despite the government having long initiated construction of nuclear power stations around the nation. Inside and outside of Fukushima, the citizens then, with the help of various grassroots initiatives, immediately started to measure radiation in their environments and food products.

JIM-NET (Japan-Iraq Medical Network) is a Japanese NGO, with a primary mission to support vulnerable Iraqi children with cancer. From the onset of the Eastern Japan Earthquake and tsunami, JIM-NET has also worked to support disaster-affected people, especially in the Fukushima area, based on the following three

M. Komatsu (✉)
JIM-NET (Japan-Iraq Medical Network), Tokyo, Japan

Fukushima Booklet Committee, Tokyo, Japan

M. Indrawan et al. (eds.), *Civic Engagement in Asia*,
https://doi.org/10.1007/978-981-16-9384-7_12

objectives: (1) to support Fukushima citizens to make informed decisions on their own future; (2) to raise awareness of the connection between Fukushima and ourselves (particularly those of us who live in Tokyo), given that the Fukushima Nuclear Power Plants were built by TEPCO to generate electricity for Tokyo; and (3) to connect Fukushima to the world, sharing lessons learned in Fukushima.

In the wake of the events of March 11, 2011, citizens' radiation measuring stations were much sought after, in order for Fukushima citizens to gain first-hand information about the level of radiation contamination in their everyday environments. In some cases, scientists and experts in radiation took the lead in building networks of "citizen scientists," whereby the local residents themselves began to conduct radiation measurements and related advocacy work. In other cases, those seeking to measure radiation were just normal citizens with no previous knowledge or expertise who wished to gain access to the methods and information. With the help of a networking NGO, from April 2012 to the present JIM-NET has incorporated into its Fukushima Project provisions for various kinds of support for the citizens' radiation measuring stations.

By the beginning of April 2012, there were already about 25 initiatives in place to conduct radiation measurements, especially for food products in Fukushima prefecture. In one such initiative, concerned mothers gathered and started to measure radiation levels. They were worried about their children's exposure to radiation in their homes and school districts, and via their intake of food and milk. In Koriyama city, located in the middle of Fukushima prefecture and with a larger population than the prefectural capital, a group called 3a! Koriyama—*Anzen* for "safe," *Anshin* for "secure," and *Action*—rented an office space and opened their measuring station in the city with funds provided by their Tokyo-based company and a few religious organizations. Other initiatives included groups of organic farmers, brokers and shopkeepers of Fukushima foodstuffs. Fukushima prefecture is a farm-rich area and had long been known as the provider of rice, vegetables, and fruits to the Tokyo metropolis and to the eastern Japan area. This nuclear power plant accident tarnished their proud reputation, causing Fukushima farm products to be simply avoided or even refused by the consumers from fear of radiation. It was indeed a dismal period for Fukushima agriculture. Only scientific evidence acquired by measurement would convince the public of the safety of their produce.

These initiatives all arose from context-specific concerns regarding radiation levels. Geographically, they were quite scattered across the three areas of Fukushima prefecture. The level of radiation contamination was quite diverse within each location. Much depended on the nature of precipitation events at the time of the initial Fukushima Daiichi release and explosions in March 2011. In our preliminary interviews, the residents requested opportunities to learn some specialized knowledge about radiation from experts. The unsettling feelings around the issues of radiation were also expressed as they discussed how to organize the measuring schemes. As a result, a study-group was organized to accommodate these two interests: a) measuring radiation and b) facilitating networking among motivated citizens. The citizens' radiation measuring stations were expected to disseminate among the local residents' scientific knowledge of radiation and fundamental know-how regarding radiation safety.

Every participant was highly motivated to learn, and the study-group approach achieved its mission quite quickly, with each measuring station equipped with the proper facilities, knowledgeable staff, and advisors. JIM-NET called for area-specific get-togethers to build networking platforms between the measuring stations that also might include some radiation experts. It appeared reassuring to physically congregate and discuss, face-to-face, how each was tackling the invisible threat of radiation. Having become acquainted in person allowed the measuring station staff to contact each other when needed for help or advice. The topics included techniques in measurement, how to interpret the acquired data on radiation levels, and ways to explain the data to lay people. Much discussion centered around the issue of management and especially about funding for it. This networking function was maintained for over a year, and it served as a platform and as a bridge between groups and sometimes with experts.

Such a network of citizens' radiation measuring stations has faded out rather quickly in Fukushima. The funding was relatively unproblematic during 2011–12, as many donors, both within and outside of Japan, were willing to support such citizen-led initiatives. That allowed the purchase of the relatively expensive measuring devices at the start-up. Human resource and running costs of the measuring activities, however, were often not covered by such funds and left each organization responsible for the financing of those activities. In our observations, we noted that those people who undertook measuring initiatives alongside a primary occupation—such as running a local grocery store, organic diner, or farm—gradually gave up maintaining the measuring stations with their own budgets. This is partly because the scientific research gradually revealed the significant decrease in, and limited amount of, radioactive material transferred from the contaminated soil to the agricultural products. Furthermore, they strongly hoped to re-establish their normalcy and get their business back on track as soon as possible. For them, radiation measuring was a tool to determine the right timing to regain control over their diminished trades.

In contrast, some groups regarded radiation measuring as their mission and the purpose of their existence. They often managed to continue their measuring activities beyond two years. Those groups who have continued measurement activities longer are the ones who set up their measuring stations as independent operations, many of which manage to continuously draw funds from outside of Fukushima. Some of the funds received are organizational donations or project funds via foundations, but some continue to come from concerned citizens all over Japan and the globe. These independent stations in Fukushima have become even more specialized to include other types of radiation measurements, such as hot spots and beta rays.[1] Some farmers and retailers also chose to continue measuring longer, at the cost of much time and effort. They continued to use their measurement data also in their advocacy

[1] For more scientific details, U.S.NRC (2018) at this site may be helpful: https://www.nrc.gov/reading-rm/basic-ref/glossary/.

or political statements, or as a counterproof to combat the negative associations of their homeland as "a forbidden, contaminated land."

More than ten years have now passed since the initial radiation fallout. The public measurement of radiation in foodstuffs has been well incorporated into the official system. Outside of Fukushima, however, public perception of Fukushima as "a dangerous land tarnished by radiation" long persisted despite a significant decrease in radiation levels and the implementation of international measurement standards. Moreover, radiation contamination was not confined to the geographical boundaries of Fukushima prefecture. With a high level of awareness, and for preventative purposes, as many as 30 citizens' radiation measuring stations have been founded outside of Fukushima. They have been actively collaborating with each other and with a Fukushima measuring station. Their measurement data have been assembled under one cooperative website for ease of public reading.[2] In 2017, the group completed the measurement of soil samples from over 3000 locations in eastern Japan, which were then compiled in map form for publication. The book was well received and won several awards in recognition of their organized efforts.

There are many more nuclear power plants in the world with inadequate, if any, disaster preparedness. Interested parties, including industries and governments, have nonetheless continued to sell new nuclear power plants to developing countries. Therefore, networks of concerned citizens will continue to be critical in ensuring nuclear disaster preparedness around the world. In order to share Fukushima's lessons with the world, the "10 Lessons from Fukushima" booklet was compiled through the efforts of another citizens' initiative in 2015. Its publication coincided with the Third UN World Conference on Disaster Risk Reduction in that year. The book is free to download (Fukushima Booklet Committee 2015), and many have volunteered to translate the booklet into various languages. The Fukushima Booklet Publication Committee is planning to continue sharing these bitter yet critically important lessons learned from Fukushima with the world, using various media and through the continued empowerment of local initiatives.

Appendix

The 10 key lessons from Fukushima[3]

What Happened at Fukushima and 10 Lessons Learned

1. Do not be fooled by the "Nuclear Power is Safe" propaganda.

[2] https://minnanods.net.

[3] This is a distillation of the contents of Chapter 2 (Fukushima Booklet Committee 2015). For Japanese language see here: http://fukushimalessons.jp and for English and other languages see here: http://fukushimalessons.jp/en-booklet.html.

2. During an emergency, the basic premise is to run away.
3. Access to information and keeping records is vital.
4. People affected by the disaster have the right to a comprehensive health survey and disclosure of information.
5. To ensure food safety and to protect agriculture, forestry, and fishery industries, citizens must participate in measurement. Information disclosure is also vital.
6. Complete decontamination is impossible.
7. The accident cannot be brought to a conclusion unless workers are given better treatment and healthcare.
8. Rebuilding the daily lives and communities of those affected is essential.
9. Those affected by the disaster should be consulted and participate in the enactment and implementation of laws that are designed to protect them
10. Taxpayers should be aware that the electrical companies depend on our tax money to cover the compensation costs.

References

Fukushima Booklet Committee. 2015. "10 Lessons from Fukushima: Reducing Risks and Protecting Communities from Nuclear Disaster." PDF, Lessons from Fukushima website. http://fukushimalessons.jp/assets/content/doc/Fukushima10Lessons_ENG.pdf.

U.S.NRC. 2018. "Glossary. United States Nuclear Regulatory Commission website." https://www.nrc.gov/reading-rm/basic-ref/glossary.html.

Chapter 13
Democracy in the Wake of the Fukushima Nuclear Disaster

Hiroko Aihara

Highlights

- Fukushima precipitated both a weariness and a preparedness among citizens toward the risks of nuclear radiation.
- A global approach to effectively dealing with the risks of nuclear radiation must include those who have experienced nuclear disasters.
- Deliberative democracy can help create a shared understanding of the problem and the development of a path forward.

Introduction

I am a native of Fukushima. On the day of the triple disaster now known as "3/11", I was working in Tokyo as the Research Administrative Secretary for members of the House of Councilors (the upper house of Japan's National Diet). While most people were evacuating Fukushima, I went in the opposite direction: I quit my job and returned to my hometown, Fukushima City, Fukushima Prefecture, which was blanketed with radiation despite being some 60 km away from the Fukushima Daiichi (No.1) Nuclear Power Plant (FD1NPP). Many will know by now that the triple disaster of 3/11 refers to the combined effects of an earthquake, a tsunami, and the resulting catastrophic nuclear accident at that plant.

I began furiously reporting on the situation, resulting in the production of more than 35 video reports, 25 online articles, and 20 magazine articles in the first year after the disaster. For many years, I wrote a weekly online column called "The Viewpoint from Fukushima" for the *Nikkei* online that covered the latest from the region.[1] I also

[1] This column continued until 2019.

H. Aihara (✉)
Japan Perspective News, Fukushima, Japan

M. Indrawan et al. (eds.), *Civic Engagement in Asia*,
https://doi.org/10.1007/978-981-16-9384-7_13

discussed the current situation in Fukushima in weekly segments on videonews.com (Japan) and have regularly contributed pieces on the tsunami and nuclear disaster that followed the earthquake of March 11, 2011 to media outlets such as "The Big Issue Japan," "Fujin no Tomo," and "Shukan Kinyobi."

Fukushima, then and Now

On March 3, 2011, a series of chain explosions of four reactors caused by a power outage of the reactor cooling systems at FD1NPP created a nuclear disaster of the same magnitude (INES Level 7)[2] as that of Chernobyl. It destroyed local daily life. In late 2011, the central government of Japan set up a budget of 32 trillion JPY for recovery efforts in the disaster area. The recovery efforts of the national government progressed slowly, however, and the situation in Fukushima remains highly challenging.

At the time of this writing residents, including my friends, my family, and me, continue to live in a state of "damage" and "disaster" in Fukushima. The number of people killed or missing in Fukushima Prefecture alone due to the combined disasters of the earthquakes, tsunami, and evacuation, ("Direct Deaths and Death Report under Missing") was 1,831 as of February 2022 (Fukushima Prefecture DRH 2022, 2). "Disaster Related Deaths," including from sickness, accidents, and suicides stood at 2,331 as of February 2022 (Fukushima Prefecture DRH 2022, 2). In Fukushima, prolonged evacuation time has led to more disaster-related deaths than direct deaths.

I continue to reside in my hometown of Fukushima City, where I have been reporting on the situation since March 2011. Through my work, I focus on public dialogue during this catastrophic situation in Fukushima, introducing local people's opinions, conversations, and feelings. My approach has been to uncover facts by interviewing native citizens. I have found that they often raise philosophical questions, such as: What is the basis for my and our happiness? What is life? How can we make a peaceful society? Are we victims or survivors? and How can we build a brighter future? These questions are very simple, but primordial. Thus, I focus on ordinary people's voices in reporting so as to describe how survivors have responded in the tragedy's aftermath.

Key Issues

Let me now highlight below a few key issues facing ordinary citizens who have been affected by the FD1NPP disaster:

---

[2] INES stands for the International Nuclear Event Scale.

Radiation Contamination

After the disaster, soil, rivers, the sea, and human bodies were contaminated with radiation that came from FD1NPP. Vegetables and fruits contaminated with radioactivity (above the legal threshold for Cesium 137 concentration) could therefore not be sold.

Decontamination Operations

Decontamination operations are carried out only in residential areas and public sites, not in the mountains, rivers, lakes, or the ocean. Contaminated waste from these operations is still piled up in the town. In Fukushima City, my hometown and where my house is located, black decontamination bags (huge plastic bags filled with contaminated waste) were often buried under residential gardens due to a lack of space and interim storage facilities in Fukushima. From an environmental justice perspective, this situation needs be exposed and rectified.

Domestic and International Evacuees as "Nuclear Refugees," "Displaced People," or "the Diaspora"

The 3/11 disaster resulted in the evacuation from Fukushima Prefecture of approximately 160,000 people altogether (Fairly 2018). Roughly half of these were ordered to evacuate from homes that were in highly contaminated areas designated official evacuation zones by the government; in their case, evacuation was obligatory. Another half or roughly 80,000—fearing for their safety—evacuated on their own accord from areas within Fukushima Prefecture that were not designated evacuation zones by the government. The evacuation from the areas around the FD1NPP took roughly two weeks and was often chaotic, with day-long traffic jams and confusion as a result of poor or crossed communications (Fairly 2018).

In February of 2018, Human Rights Now reported that the Japanese government had been lifting evacuation orders for some official evacuation zones after changing the amount of allowable public exposure to radiation in those areas—from the international standard of 1 mSV/year to the much higher level of 20 mSv/year (HRN 2018).[3] In line with these decisions, the government of Fukushima Prefecture announced in 2017 a timetable for withdrawing housing support to evacuees from areas now declared as "cleared" under the new standards (HRN 2018). In March of 2017, the national government ended housing support for the 32,000 voluntary

[3] This decision has come under criticism from a number of quarters, including Japanese organizations such as the Citizens Nuclear Information Center (CNIC 2017) and from the U.N. (Gunter 2018).

evacuees still displaced at the time, and then dropped them from the lists of evacuees, rapidly and artificially reducing the number of the displaced (HRN 2018). Naturally, such decisions have created economic pressure to return for people who may not feel it is safe to do so (CNIC 2017). By September 2019, roughly 30% of the official evacuation zones had been declared cleared (339 sq. kms. of the total 1,150 sq. km. area), excluding severely contaminated areas (Reconstruction Agency Japan 2019). As the number of cleared zones expand, larger numbers of evacuees will face the pressure to return under uncertain and potentially unsafe conditions. The rate of return has in fact varied regionally due to continued radiation contamination and the lack of social capital and infrastructure (transportation, supermarkets, hospitals, schools, neighborhood community facilities, and so on).

As of March 2020, nine years after the disaster, the Fukushima Prefecture Government Office reported that 40,335 people who originally resided in Fukushima Prefecture were still living as evacuees (Fukushima Prefecture DRH 2020, 7). Of the total number of those counted as evacuees in March 2020, 9,408 had evacuated to other areas within Fukushima Prefecture, while 30,914 had evacuated to areas outside of Fukushima; the whereabouts of another 19 were unknown (Fukushima Prefecture DRH 2020, 1). The same government office reported that 16,979 of those who had evacuated to areas outside of Fukushima were still residing in temporary evacuation houses, 13,749 people were staying in the house of a relative or acquaintance, and 186 were in hospital (Fukushima Prefecture DRH 2020, 7).

Mistrust and Controversy Regarding Losses and Damage to Health Experienced by Residents of Fukushima

No one can deny that the triple disaster of 3/11 brought about tremendous human suffering that included, in the words of one writer, "evacuations, emotional trauma and premature deaths, disrupted jobs and schooling" (Normile 2016). Still, in one area the findings have been controversial despite massive testing, namely the incidence of thyroid cancer in children. In general, the damage to human beings from radiation is difficult to measure precisely. But based on the appearance of thyroid cancer in children after the Chernobyl disaster, the Fukushima Prefectural Government had good reason to carry out massive tests that aimed to screen all young people—from newborns to 18-year-olds at the time of the disaster—for signs of thyroid abnormalities and other health indicators. The tests for thyroid abnormalities were given in several rounds to establish data over time, and the entire operation was supported by the national government. The number of individuals in the target age group for the first screening beginning October 2011 was roughly 368,000, of whom about 80% were successfully tested (Yamashita et al. 2018). Of these, 116 showed signs of malignant or suspicious lesions, and 101 were declared to be cancerous following thyroid surgery (Yamashita et al. 2018). Furthermore, about half of these children showed abnormalities of the thyroid that led many to the conviction that these were a result of 3/11 (Normile 2016).

As of 2016, the scientific community seemed to agree that these thyroid cancer cases and abnormalities could *not* be attributed to 3/11, based on a number of reasons. The first is the "screening effect," which argues that very large-scale screenings will uncover more cases than would normally appear through clinical visits upon showing symptoms (Rosen 2020, 1). Second, the screenings relied on a relatively new ultrasound technology that was able to show much smaller abnormalities or lesions than the older technologies (Normile 2016). Third is that the onset of thyroid cancer does not normally occur until several years later. Fourth is that evacuation happened relatively rapidly, and much of the initial release of radiation was swept out into and over the Pacific Ocean. Finally, tests in other regions of Japan using the same ultrasound technology showed roughly similar levels of abnormalities and have led to surprise about such high levels in children generally (Normile 2016).

However, a report in the March 2020 newsletter of the German affiliate of International Physicians for the Prevention of Nuclear War (IPPNW) raised serious questions about the earlier scientific consensus. Author of the report, Dr. Alex Rosen, made several claims. Based on four rounds of screening from 2011 to the present, by February 2020 the total number of cases of thyroid cancer among children in Fukushima came to 197 and suspected cases to 50, for a combined total of 247 cases of verified or suspected cases of thyroid cancer. 197 verified cases amounts to a rate 17 times greater than the expected rate of childhood cancer in Japan based on surveys done from 2000 to 2009 (Rosen 2020, 2). The screening effect cannot be blamed, as 96 of the verified cases appeared in follow-up tests of the population already screened, many of whom showed no abnormalities in earlier screening (Rosen 2020, 1). Rosen also noted a pattern correlating higher rates of thyroid cancer with closeness to the FD1NPP (Rosen 2020, 3).

Rosen observes finally that participation in successive screenings dropped steadily, potentially underreporting the actual incidence of thyroid cancer in children. This drop appears to have been encouraged by medical examinations that include a choice to opt out of future screenings, and by Fukushima Medical University staff who regularly visited schools explaining the right "not to know." Rosen asks if these decisions might in any way reflect a larger political agenda of the Japanese government or the IAEA[4] to downplay the health dangers of nuclear power plants and the 3/11 disaster (Rosen 2020, 5). A *Guardian* writer echoes this concern, commenting that "The anxiety felt by parents in Fukushima stems from a widespread lack of trust in the local medical authorities, which have come under government pressure not to cause alarm among residents" (McCurry 2014). In the midst of the debate and the confusion on these questions throughout the period 2011 to the present have been parents and other residents who feared for their health and the health of their children. What should they now believe? What should they conclude? What should they imagine about the future?

---

[4] International Atomic Energy Agency.

Strategy—Toward a Global Network of Nuclear-Affected Persons Through the Practice of Deliberative Democracy

Nine years have passed since 3/11. The Fukushima situation is a Glocal (global and local) issue in the nuclear age. As a disaster frontline journalist in Fukushima, I would like to share the lessons I have learned from this human-made nuclear disaster. In this nuclear age, more nuclear disasters will occur elsewhere in the world. This means that our desperate experiences in Fukushima are useful not only for the survivors in that locality, but for everyone who wishes to prevent, or at least diminish the impacts of, such future disasters.

Unfortunately, year by year, fewer and fewer journalists report on Fukushima. This is due to the lack of awareness on the part of the mass media. In order to overcome this, we need to present new perspectives on the future from all nuclear-affected persons (*hibakusha*).[5] We can communicate and open spaces for discussion with other nuclear survivors in or near Hiroshima, Nagasaki, and Chernobyl, as well as with downwinders—individuals and communities exposed to radioactive contamination or nuclear fallout and nuclear accidents—from nuclear test sites and uranium mines in the western US and Bikini-Marshall Island. Doing all of this will be tantamount to creating a Global Hibakusha Network.

I have worked primarily with Fukushima residents, media representatives, journalists, and students. In 2013, I started a research project with my colleagues and students, sponsored by the Toyota Foundation, to create "The Global Hibakusha Network." In this project, a diverse group of people has come together. They are people concerned about Fukushima after the disaster and nuclear power plant explosion—Fukushima university students, researchers, journalists, and a movie director. To date, we have carried out the activities I outline below, many of which can be considered examples of "deliberative democracy."

Deliberative democracy is a form of democratic participation that goes far beyond voting as a public expression of preferences. Central to the idea in its early phases was that it would involve a kind of democratic public discussion that would invite people to consider the common good together using rational arguments. It thus imagined a different kind of politics than one that is merely a public space for negotiating conflicting and partial interests. Dryzek and Niemeyer offer a broader view of what counts as deliberative democracy in their definition from 2012:

> Our view is that we should allow pretty much any kind of communication that is non-coercive, capable of inducing reflection, strives to link personal viewpoints to larger principles, and tries to make sense to others who do not share the speaker's framework. So in this light we can (conditionally) accept the telling of personal stories, rhetoric, humour, ceremonial speech, even gossip, as well as arguments. Threats, lies, abuse, and command have no place. (Dryzek and Niemeyer 2012)

[5] The Japanese term "hibakusha" originally referred to the victims of the atomic bombs dropped on Hiroshima and Nagasaki on August 6 and 9, respectively, in the year 1945. More recently, the term has come to be applied much more widely to include all those affected by exposure to nuclear radiation.

I believe the activities of the Global Hibakusha Network in their many forms encompass this broader sense of deliberative democracy. Among the activities we have undertaken are the following:

- We coordinated discussions among various stakeholders, including government officials, policymakers, media representatives, journalists, academics, teachers, civil society organizations, and citizens.
- We carried out "2D" (debate and discussion) conversations and focused on deliberative discussion involving stakeholders from various sectors. These discussions establish common understanding of the problems. Deliberative democracy follows the methods of deliberative conversations, such as at the consensus conference (in the UK, the Netherlands, New Zealand, and Switzerland), the scenario workshop (Denmark), Future Search (US), and Citizens' Panel (the UK).
- We led participatory exercises for naming (to name our problems and difficulties), flaming (to clearly decide the discussion point as well as the scope of the problem), discussion, and activities (exchanging ideas with other people and groups).
- We facilitated mutual interviews for practical study about nuclear disasters in Fukushima, the Marshall Islands, and other areas with our group members and meeting participants. This helped build a new network of wisdom among survivors. Since 2013, Fukushima and Waseda University students and our research group members (Peace Studies Association of Japan) visited the Marshall Islands and Fukushima to have a deliberative conversation. In 2014 and 2015, we invited several resource persons to discuss our future in a post-3/11 society. These included a survivor of the nuclear testing on the Ailuk Atoll and a former national council member of the Marshall Islands.
- We have continued collaborative activities with various organizations by hosting fieldwork and visits in Fukushima to share our experiences with university faculty members, students, and youth (so far from Meisei, Ibaraki, Waseda, and Meiji Gakuin universities), scientists, experts, media representatives, and journalists. We used not only traditional media, but also Facebook, Twitter, and other social media, to reach out to writers and reporters who wanted to enrich their experience and expertise. We continue to pursue deliberative democratic collaboration with various stakeholders, including policymakers.
- We supported collaboration and solidarity between civic societies in Fukushima, Hiroshima, Nagasaki, Chernobyl, and the Marshall Islands, including activities such as mutual interviews and study (inviting speakers from the Marshall Islands, Hiroshima, and Nagasaki), a field trip to the Marshall Islands (on March 1, Memorial Ceremony Week), and open seminars (in Fukushima, Hiroshima, Nagasaki, Minamata, and elsewhere).
- Lastly, in order to build Civic Archives in Fukushima, we collaborated with the documentary filmmaker Mr. Yasuyuki Mori and launched a new film entitled *Tanemaki Usagi* (Snow Rabbit) about planting rice after 3/11. The film is one of the tools we subsequently utilized to start conversation and discussion toward mutual understanding about Fukushima, the Marshall Islands, Yaizu City in Shizuoka,

and the Lucky Dragon fishing boat exposure to atomic bomb testing by the US on the Bikini Atoll in 1954.

I have started building a network in Fukushima with citizen scientists, NGOs, and universities. We also created a network map that includes the non-profit organization center "Peace Museum of Auschwitz" and "Nuclear Power Plant Disaster Information Center," which were built by individual citizens in Shirakawa City, Fukushima. Our activities provide open resources such as videos, interviews, and information. Through sharing, we have created a new realm—from the perspective of Global Hibakusha and Peace Studies—within which to talk and learn. We interview each other about nuclear disaster and damage and redefine nuclear disaster through objective feedback, verbalized concepts, thoughts, and descriptions. We strongly encourage the young generation to participate and to have an open mindset.

Paths Forward

The safety of living near nuclear power plants has become a very hot topic. Using a deliberative democratic approach, survivors of the 2011 Fukushima disaster have begun to chart a roadmap to recover from the tragedy. They have also used the approach to open dialogues with other communities around the world that have suffered the impacts of nuclear radiation and fallout in order to develop a global approach to mitigating and building resiliency to the risks posed by nuclear radiation.

Meanwhile, the Japanese government is now trying to export nuclear technologies and power plants by concluding nuclear energy agreements with other Asian countries. Thus, we remain under the crisis of nuclear proliferation. The FD1NPP disaster has largely faded from national and global consciousness and is now considered by many to be an isolated problem only affecting Fukushima and nearby communities. It is critical, however, that we learn lessons from Fukushima and Global Hibakusha experiences in order to build a sustainable society.

In doing this, we face continued difficulties. The first is how to transmit our experiences to younger generations who may not have strong recollections of the incident and whose attention is often devoted to other issues such as pursuing their education and career. The second challenge is that activities within the civic sector do not often translate into legislation and policymaking; there is a need for more efforts to expand the conversation among stakeholders and policymakers to improve the relevant governmental policy frameworks. We have taken a step in this direction by inviting local government officers, schoolteachers, and local assembly members to improve government policy. We must create further avenues through which stakeholders can inform and improve legislation.

Lastly, there is a limitation of resources. Sometimes, we are forced to confront the lack of resources, money, and facilities. We are currently exploring research funds as one possible solution. We also hope to expand the circle of discussion and learning; we would like to develop exchanges with other non-nuclear disaster sites in the ASEAN region, such as the tsunami-affected area in Aceh, Indonesia, about ways to respond to disasters.

Radioactive Waste Found at Temporary Storage Site in Naraha Town, Fukushima Prefecture, Japan, July, 2018.

Evacuees From The Atolls, Japanese Students, and Researchers Gather During Fieldwork in the Marshall Islands, March 2014.

References

CNIC (Citizens' Nuclear Information Center). 2017. "Fukushima Evacuees Abandoned by the Government." CNIC website. March 31, 2017 (updated from the original publication date of August 2, 2016).

Dryzek, John, and Simon Niemeyer. 2012. "What Is Deliberative Democracy?" Centre for Deliberative Democracy and Global Governance (Australian National University) weblog. February 15, 2012.

Fairly, Ian. 2018. "Fleeing from Fukushima: A Nuclear Evacuation Reality Check." Beyond Nuclear International website. March 11, 2018. https://beyondnuclearinternational.org/2018/03/11/fleeing-from-fukushima-a-nuclear-evacuation-reality-check/.

Fukushima Prefecture DRH (Fukushima Prefecture Disaster Response Headquarters). 2022. "Updated Report on the Damage Situation Resulting from the 2011 Earthquake Off the Pacific Coast of the Tohoku Region, No. 1763." February 2, 2022 (Japanese language source).

Gunter, Linda Pentz. 2018. "Fukushima Evacuees Forced Back into Unacceptably High Radiation Zones." Beyond Nuclear International website. December 2, 2018. https://beyondnuclearinternational.org/2018/12/02/fukushima-evacuees-forced-back-into-unacceptably-high-radiation-zones/.

HRN (Human Rights Now). 2018. "Evacuees of the Fukushima Nuclear Disaster Will Lose Government Assistance Soon" (Written Statement Submitted to 37th Human Rights Council Session). Human Rights Now website. February 5, 2018. http://hrn.or.jp/eng/news/2018/02/05/hrc37fukushima/.

McCurry, Justin. 2014. "Fukushima's Children at Centre of Debate Over Rates of Thyroid Cancer." *The Guardian*, March 9, 2014. https://www.theguardian.com/world/2014/mar/09/fukushima-children-debate-thyroid-cancer-japan-disaster-nuclear-radiation.

Normile, Dennis. 2016. "Mystery Cancers Are Cropping Up in Children in Aftermath of Fukushima." *Science* website. March 4, 2016. https://www.sciencemag.org/news/2016/03/mystery-cancers-are-cropping-children-aftermath-fukushima.

Reconstruction Agency Japan. 2019. "Fukushima Prefecture Evacuation Order Zone Situation" (Japanese language source). https://www.reconstruction.go.jp/topics/main-cat1/sub-cat1-4/fuhyou/pamphlet/saishin2P_20190920.pdf.

Rosen, Alex. 2020. "Thyroid Cancer in Fukushima: 9 Years after the Multiple Nuclear Meltdowns." In *IPPNW Information*, March 2020. https://www.ippnw.de/commonFiles/pdfs/Atomenergie/Fukushima/IPPNW_info_thyroid_cancer.pdf.

Yamashita, Shunichi, and Shinichi Suzuki, Satoru Suzuki, Hiroki Shimura, and Vladimir Saenko. 2018. "Lessons from Fukushima: Latest Findings of Thyroid Cancer After the Fukushima Nuclear Power Plant Accident." *Thyroid* 28 (1), published online January 1, 2018. https://www.ncbi.nlm.nih.gov/pmc/articles/PMC5770131/.

Chapter 14
Fighting Modern Slavery in Southeast Asian Waters

Sompong Srakaew and Patima Tungpuchayakul

Highlights

- Modern human slavery in the form of forced and abusive labor in the fishery sector continues to occur in Southeast Asian waters.
- The victims may lose everything, including their memories, identities, limbs, and lives.
- The fight against modern slavery requires long-term persistence and support from all the stakeholder groups across borders.

Introduction

Migrant workers are becoming the center of attention in Thailand. Statistics compiled by the Kingdom of Thailand's Department of Employment and Office of Foreign Workers Administration, a subsidiary of the Ministry of Labour, show that the number of migrants working in Thailand and Southeast Asia, both legally and illegally, has risen consistently in recent decades.[1] The formation of the ASEAN Economic Community (AEC) and its associated Free Trade Agreements are expected to further spur an increase in the movement of ASEAN citizens seeking employment. Consequently, Thailand is expected to experience both an influx and an outflux of skilled as well as unskilled labor.

[1] The Thai language website of the Foreign Workers Administration Office, under the Department of Employment of the Thai Ministry of Labour, offers extensive statistics on a monthly and yearly basis on foreign workers working in Thailand and on ASEAN citizens working in Thailand, going back in some categories as far as 2001. See https://www.doe.go.th/prd/alien/statistic/param/site/152/cat/82/sub/0/pull/category/view/list-label [Accessed April 13, 2020].

S. Srakaew (✉) · P. Tungpuchayakul
Labor Rights Promotion Network Foundation (LPN), Samut Sakhon, Thailand

M. Indrawan et al. (eds.), *Civic Engagement in Asia*,
https://doi.org/10.1007/978-981-16-9384-7_14

Such an increase in the movement of people without corresponding protection mechanisms is likely to lead to an increase in human rights issues. As migration rates rise, more people are made vulnerable to employer abuse, labor exploitation, and human trafficking. If Thailand does not urgently address the issues of forced labor and exploitation, these problems will only become more widespread as illegal networks and trafficking rings continue to operate with impunity.

This chapter chronicles how regional NGO activists have fought slavery and rights abuses in the region's waters.

Human Trafficking and Modern Slavery in Ambon and Surrounding Islands

Human trafficking and slavery have increased in the region over the years. Ambon and the surrounding islands of the Maluku archipelago in Indonesia provide an illustrative example.

Thai fishing vessels have been operating in international waters, especially off the Indonesian coasts, from 1965 onward. In 2006, the numbers of these vessels increased to around 1,500 (Tangprachyakul 2015, 1).[2] These fishing operations have led to forced labor, slave labor, detention, and severe physical abuses of fishing workers, especially those victimized by human trafficking syndicates. In many cases, laborers were tricked into enslavement on boats for periods of six years or more. Once at sea, they often find it impossible to get home and may end up stranded on small islands for many years. Fraudulent documentation is common, making it impossible for workers to maintain identity papers (Tangprachyakul 2015, 1).

Laborers on these boats mainly come from Thailand, Myanmar, Laos, and Cambodia. Some are mere boys, on average 12-year-olds. Mr. Saksit, an orphan from the age of six, was kidnapped and enslaved on the fishing boats. Even after being rescued, he struggled to uncover his true identity (personal communication).

The modus operandi of such operations and the suffering they caused is now gradually being uncovered. In Thailand, the problem began with agents and some groups claiming to be governmental staff arresting and detaining the migrant workers from Myanmar and Cambodia. The workers were then sold to agents to work on fishing boats, primarily for Thai employers. The kidnappers are usually brokers from neighboring countries who turn a profit sending the fishermen to Thai employers (Tangprachyakul 2015, 2).

---

[2] Our work has involved extensive field research and ongoing work on the ground, resulting in numerous articles, videos, and a documentary film. English language materials can be accessed via our site at https://www.lpnfoundation.org, Thai language materials at https://lpnthailand.org. Much of the background information we provide in this section relies on a key summary document in English authored by our co-founder, Patima Tangprachyakul (2015) and available at https://lpnthailand.files.wordpress.com/2015/08/the-special-report-on-operations-saving-workers-in-fishing-boats-from-indonesia.pdf.

Confronting the Problem

Something had to be done to save workers from being enslaved on fishing boats across Southeast Asian waters. This became the mission of the Labour Rights Promotion Network Foundation (LPN), with its two operational arms, the Seafarers Action Center and the Labour Center (Tangprachyakul 2015).

LPN's regular investigations uncovered many startling findings, especially from Ambon and the surrounding islands. For instance, LPN's field investigations between 2006 and 2010 suggested that more than 70% of the migrant fishermen[3] working in Indonesia had not received fair wages or payment, and most of them were forced into slavery to satisfy their supervisors and employers (Tangprachyakul 2015, 2). Further, Myanmar workers were regularly sold to fishing boats (especially during the period 2006–2010) and were illegally registered as Thai workers in fraudulent seamen books. In some cases, migrant workers had not returned home for more than 22 years (Tangprachyakul 2015, 2). Most workers could not return home because they did not have any documents or identity papers, which had been confiscated and kept with the boat captains. This included identity fraud by Thai and Indonesian fishing companies that engaged in a corrupt system to change the nationalities of the workers in order to camouflage working relations as well as family ties, so that, respectively, employers could avoid responsibilities, and the family could not track the missing person. The LPN team also documented more than 500 Thais who were drugged and forced into boats; some of these were delirious and suicidal upon their return (Tangprachyakul 2015, 2).

LPN does its best to visit, follow up with, and provide assistance to Thai and migrant fishers who are trafficked and have experienced exploitative conditions. LPN continues to update the status, problems, and needs of Thai and migrant fishers, and provide humanitarian assistance. These activities are conducted together with concerned parties and stakeholders from Indonesia and Thailand.[4] Victims also sometimes participate in rescue operations. For example, one rescued Burmese fisherman returned to Ambon Island where he volunteers to help find others stranded for years.

In 2006, 66 fishermen from the boat *Prapat Navee*, which had been operating in Indonesian waters, asked LPN for assistance. Out of the sixty-six personnel, 39 (Thai and Burmese) passed away before or during their return journey to their motherlands. Some were seriously ill and needed treatment and care. None of the fishermen received any payment for their work or compensation for their ordeal (Tangprachyakul 2015, 1).

---

[3] Migrants from Myanmar, Cambodia, Laos, and Thailand working on Thai boats in Indonesian waters.

[4] The rescue missions were a collaboration between LPN, village/community heads in Benjina Island, and local immigration and civil society representatives on Ambon Island. In Thailand, LPN collaborated with The Mirror Foundation, Stella Maris, the Human Rights and Development Foundation (HRDF), UN-Act, and the Alliance Anti-Trafic (AAT). In the follow-up work, LPN cooperated with local authorities, namely village/community heads on Ambon Island.

During 2006 to 2014, the LPN Foundation responded to 128 complaints of fishermen missing in Indonesian waters. Over the course of the eight years, LPN discovered no less than 39 deaths and two injuries that led to permanent disabilities. Survivors never received any care from the syndicated employers, let alone proper compensation for their hard work under extremely harsh living conditions (Tangprachyakul 2015, 2).

LPN has conducted rescue operations of migrants from different Southeast Asian countries in Ambon and surrounding islands (Tables 14.1 and 14.2).

Table 14.1 LPN's Rescue Operations between August 2014 and August 2015

	Dates	Activities	Result
1	August 19–24, 2014	– First survey found 6 Thai fishermen asking for help, built temporary shelter and provided food and accommodation for the victims, coordinated with the government counterparts (the DSI—Department of Special Investigation), Indonesian Immigration, the Thai Consulate in Ambon), and INGO counterparts (IOM, UNACT)	The first group of Thai victims rescued and returned to Thailand on October 1, 2014
2	September 11–23, 2014	– The first group of Thai victims rescued and returned to Thailand on October 1, 2014	15 fishermen rescued
3	November 15–December 3, 2014	– LPN did another survey from Ambon to Benjina, found vast graves of Thai fishermen and some victims stranded on the island for more than 18 years	10 fishermen rescued
4	January 11–20, 2015	– LPN coordinated with the Ambon Immigration office to help child victims and facilitate the repatriation of mentally ill victims	6 children rescued 12 Thai fishermen rescued

(continued)

Table 14.1 (continued)

	Dates	Activities	Result
5	March 12–April 6, 2015	– Found 60 stranded fishermen on Ambon island, but could only rescue 10 at a time because of limited transportation facilities.[a] – Media coverage on the mass graves, victims, and detained victims, resulting in the return of fishermen on a C130 flight – Monitoring on Ambon and Benjina islands – Indonesian government ordered its navy boats to rescue and repatriate 500 fishermen (Laotian, Burmese and Cambodian) from Benjina island	21 fishermen rescued from Ambon island 68 Thai fishermen rescued and repatriated by C130 flight 5 detained fishermen returned home
6	March 12–April 6, 2015	– LPN returned to Indonesia to provide assistance to the fishermen in Indonesia; 700 more Thai fishermen returned home from Benjina island	Fishermen gradually arrived in Thailand, but did not receive proper labor rights protection from the government
7	August 20–31, 2015	– DNA testing and body exhumation – 500 Burmese fishermen still waiting to be repatriated from Ambon Island; three have lost eyesight and hands from work but have not been compensated. All workers have not received wages, ranging from 1–7 years – 500 fishermen left stranded on the Ambon, Benjina, and Tual islands. Some fell sick and expressed the need to return home to restore their health but have not received assistance from the employers	

Table 14.2 Nationalities of Migrants Rescued in LPN Operations during August 2014 to August 2015

Nationality	Number	Coordinating Agencies	Remarks
Thai	1,613	Ministry of Foreign Affairs (Thailand) + Business Owners (Thailand) + Paveena Foundation + LPN	Distributed over Maluku islands (Ambon, Benjina, Tual) and Pontianak/ Kalimantan
Burmese	628 (estimated)	Indonesian government + IOM	300 fishermen were waiting for nationality verification on Ambon island LPN coordinated with stakeholders to resolve the issue of unpaid wages[a]
Cambodian	65	Indonesian government + IOM	LPN coordinated with stakeholders to resolve the issue of unpaid wages[b]
Laotian	14	Indonesian government + IOM	LPN coordinated with the stakeholders to resolve the issue of unpaid wages[c] LPN contacted victims' families as a crucial part of repatriation

Multipronged Approaches to Assist Migrant Workers

Over the past 15 years, LPN has worked continuously with migrant workers and migrant children. This work has resulted in the development of various types of strategies and methods, aimed at improving the working and living conditions of migrant workers. These include:

1. **Collaboration:** LPN collaborates with other activists and official bodies in order to deliver the most effective assistance and treatment to victims of human trafficking and human rights abuse. Examples of this activity include the establishment of the Labour Centre and the Seafarers Action Centre in Thailand, both of which work in cooperation with law enforcement groups.
2. **Lobbying:** LPN regularly engages with Thailand's Ministry of Social Development and Human Security, the Ministry of Labour, and the Ministry of Foreign Affairs in order to formulate ways to manage and administer Thailand's foreign labor practices and work toward introducing more effective preventative measures to combat human trafficking. This collaboration has resulted in an official agreement between these government ministries and Thailand's labor rights activists, which serves to ensure that preventative steps are taken to eliminate the trafficking of persons in nine seriously affected provinces around the Gulf of Thailand. Further, LPN is involved in a new subcommittee dedicated to producing a report on child labor. The foundation is also a member of a committee within the Ministry of Labour that seeks to resolve issues of child

labor and forced labor in the shrimp fishing industry, the seafood processing industry, and the agricultural sector.

3. **Service Delivery:** As well as approaching issues of child labor and migrant discrimination from a preventative angle, LPN also strives to improve the living conditions of migrant families and works to provide assistance for those who may not have access to education or other services due to their migrant status. LPN works with a number of schools in Samut Sakhon, Ratchaburi, and Samut Prakan provinces to specifically cater to the needs of migrant children, especially from Myanmar and Cambodia. This work is done in collaboration with the International Labour Organization, Save the Children UK, the Embassy of Japan and the Embassy of the United States, the United Nations Action for Cooperation against Trafficking, Oxfam, Terre Des Hommes, and from the private sector with the Thailand Frozen Food Association, Thai Union Frozen Product PCL, the Thai Fisheries Producers Coalition, Narong Seafood Company Ltd., and the Thai Fishery Producers Coalition (TFPC).
4. **Coalition Forming:** LPN regularly encourages participation from other migrant and labor rights non-governmental organizations, such as the Action Network for Migrants (ANM), the Migrant Working Group (MWG), the Anti-Human Trafficking Network in Thailand (ATN), and the Cambodia and Thailand Anti-Human Trafficking Network (CAHT), in order to push for the establishment of the Migrant Union Network in Thailand (MUNT). This initiative also includes migrant networks, for example, the Myanmar Labour Group (MLG) and the Cambodian Labour Group (CLG).
5. **Networking:** LPN continues to engage with strategic hubs such as the Anti-Human Trafficking Network in Thailand (ATN), Cambodia and Thailand Anti-Human Trafficking (CAHT), the Thai and Migrant Fishers Union Group (TMFG), the Solidarity Committee for the Protection of Myanmar Migrant Workers (SCPM), the Multi-stakeholder Initiative for Accountable Supply Chain of Thai Fisheries (MAST), and the Migrant Working Group (MWG).
6. **Evaluation and Review:** LPN initiates adjustments to existing operations and strategies in order to develop effective and cooperative relationships with national and regional organizations and institutions, both governmental and privately run. These activities stem from an internal focus on developing best practices and are aimed at promoting equal opportunities for all people, regardless of religion, gender, nationality, ethnic background, or age.
7. **Research:** At the present time, LPN operates as a research and learning center, which caters to students at all levels of education, as well as organizations working in Thailand, Southeast Asia, and worldwide. The foundation also functions as a base for national and international media associations, distributing information about the urgency and significance of migrant labor rights and the anti-human trafficking movement.
8. **Knowledge Sharing:** LPN creates and improves methods of disseminating knowledge and information regarding the rights and legal entitlements of migrants. For example, in areas with a high-density migrant population, such as Samut Sakhon, Samut Prakan, Trat, Rayong, Bangkok, Ranong, Pang Nga, and

Kanchanaburi provinces, LPN works with local schools to ensure the inclusion of migrant issues and rights in the curriculum. This has proven to be an effective model for the integration of migrant children into Thai public schools.

LPN has also found that social media works in providing relevant information to migrant workers. Surachai Mintun, a migrant worker from Myanmar who for years worked as a fisherman, is now a champion campaigner for the cause. Mr. Mintun found that Facebook provided the best channel to communicate with migrant laborers, and he now regularly updates information about laborer's rights, migrant children's education, government policies, legal consultation, and livelihoods of migrant workers.

About LPN

The Labor Rights Promotion Network Foundation (LPN) is committed to protecting and improving the lives of migrant workers in Thailand and ASEAN. We strive to advocate for equality, both in the workplace and in the community, and aim to assist migrants and their families to integrate peacefully into Thai society.
LPN's goals are:

1. To develop an understanding of the issues faced by migrant workers;
2. To support the protection of individual migrant workers and their rights, as well as the integration of migrants and their families into Thai society;
3. To assist migrant workers in becoming more self-reliant with the support of public and private sectors;
4. To create awareness about the responsibilities toward, and the well-being of, migrants; and
5. To continue operating on a not-for-profit basis with financial assistance from the public.

LPN's areas of focus are (1) Access to basic human rights; (2) Education; (3) Health; (4) Child protection; (5) Workers' rights; (6) Prevention of child labor; (7) Prevention of human trafficking; and (8) Reproductive health.

LPN's key activities and services include (1) Assistance for victims; (2) Development of welfare services; (3) Preparation of migrant children for schooling; (4) Support for migrant children enrolled in the government school system; (5) Learning Centre for children and adults; (6) Counseling Centre; (7) Multicultural Centre for migrants; (8) Temporary Shelter; (9) Labour Centre; and (10) Seafarers Action Centre.

References

Foreign Workers Administration Office. n.d. "Information and Statistics on Foreign Workers." Foreign Workers Administration Office website. Accessed April 13, 2020. https://www.doe.go.th/prd/alien/statistic/param/site/152/cat/82/sub/0/pull/category/view/list-label.

Labor Rights Protection Network Foundation website. English: https://www.lpnfoundation.org. Thai: https://lpnthailand.org.

Tangprachyakul, P. 2015. "The Special Report on Operations Saving Workers in Fishing Boats from Indonesia, Seafarers Action Center (SAC), Labor Rights Promotion Network Foundation (LPN)." Labor Rights Promotion Network Foundation. https://lpnthailand.files.wordpress.com/2015/08/the-special-report-on-operations-saving-workers-in-fishing-boats-from-indonesia.pdf.

Chapter 15
Female-Driven Climate and Environmental Action: Champions from Pakistan

Areej Riaz and Mairi Dupar

Highlights

- Women harness knowledge, expertise, and unique experiences for fighting climate change. They therefore present an untapped asset in coping with environmental issues and inculcating their pro-environmental values in their communities and future generations.
- In global framings to date, the emphasis has too often been on women and girls as victims of climate change impacts, whereas their instrumental roles as natural resource managers, particularly in rural areas (IPCC 2019), and consumers and caregivers everywhere position women at the vanguard of championing and adopting sustainable behaviors.
- Women should be given access to relevant education, trainings, finance, and technologies to enhance, expand, and further develop their role as the protectors of the environment.
- Women should be better integrated into policymaking and planning on environmental action, because policies that are blind to women's realities—whether climate adaptation or mitigation-focused or both—are also doomed to underperform. Women can formulate and sense-check policies and are strongly positioned to link policy and planning with community and household-level implementation.

Introduction

Women play a critical role in sustainable development—from agriculture and sustainable production, to education and addressing climate change. Women are at the

A. Riaz (✉) · M. Dupar
Overseas Development Institute, London, UK

M. Dupar
e-mail: m.dupar@odi.org.uk

M. Indrawan et al. (eds.), *Civic Engagement in Asia*,
https://doi.org/10.1007/978-981-16-9384-7_15

heart of many decisions that influence environmental and climate action, as leaders, policymakers, negotiators, farmers, foresters, scientists, household managers, and consumers.

At the same time, there is undeniable evidence that women are most affected by climate and environmental problems (UN/ADPC 2010). Furthermore, women and children are far more likely to die than men during extreme weather events linked to climate change owing to their social roles and more impoverished status that together lead to greater vulnerability (Peterson 2007). In major past disasters the likes of Hurricane Mitch and Katrina, or the European heat waves, more women and children have died than men, as women are mostly domesticated, have mobility restrictions imposed by their patriarchal families or societies, and are responsible to provide care to family members young and old.

We are at a pivotal moment in the international framing of climate solutions in which government leaders, donors, and development agencies are recognizing that gender-blind climate policies and programs are not effective. The unequal status of women in all societies and the leadership of women and gender-aware men in global development processes have all led to a flurry of initiatives to acknowledge the importance of including women fully in climate action.

After all, living with climate change and preventing further global warming are development issues. Climate policies build on decades of development initiatives to improve women's well-being and they benefit from the interventions of UN Women and other specialized agencies[1] set up to tackle discrimination. It is not, then, surprising, that intentional efforts have been made to include and mobilize women in the quest for a net-zero carbon, resilient future.

The last two Executive Secretaries of the UN Framework Convention on Climate Change (UNFCCC) itself have been Latin American women; and the UNFCCC has recently concluded its major Lima Work Program (UNFCCC, n.d.) to strengthen women's role in its decision-making processes. Women are recognized in the Paris Agreement as being key to the success of adaptation initiatives, listed with other "marginalized" groups such as indigenous people (UNFCCC 2015). Women are explicitly mentioned as a key domestic constituency for implementing climate actions in 50% of formal plans that countries have submitted to the UNFCCC—the Nationally Determined Contributions—although arguably, that is 50% too few NDCs to be "gender responsive" (Gerretsen 2018).[2]

Furthermore, major international public funders of climate adaptation and mitigation, such as the Climate Investment Funds of the World Bank and IMF, the Green Climate Fund and the Adaptation Fund have all developed detailed and sophisticated guidelines on the assessment of gender-differentiated needs at the climate project preparation stage. Robust gender and social assessments are a prerequisite for funding

[1] Such as the United Nations Commission on the Status of Women (https://www.unwomen.org/en/csw) and its various national counterparts.

[2] "In a sample of 57 climate change action plans submitted for the Paris Agreement to curb global warming, researchers found that keywords such as 'poverty', 'women' and 'ethnic' did not appear once in almost half [of them]" (Gerretsen 2018).

approval. Gender specialists have been developing training manuals and tool kits on gender-responsive climate actions for more than ten years (Aguilar Revelo 2009). Being more inclusive of women in climate-related decision-making processes and climate-orientated activities, from the global to the national and sub-national level, has rightly been seen as the correct and fair thing to do.

However, we are now at a point at which women's inclusion is not only seen as correct and fair. Women's leadership is now also being acknowledged as fundamental to successful environmental outcomes. Will this dawning recognition bring a sea change in how women's talents are regarded and welcomed—nationally and locally?

A critical mass of evidence is building up concerning the way that climate policies and programs underperform when women's needs and concerns are not taken into account and when women do not have a meaningful say in decision-making. The sustained involvement of women (not just lip service at the beginning of a project or program) is now recognized as vital. For instance, CDKN (Climate and Development Knowledge Network) investigated the performance of diverse climate adaptation initiatives across three continents: in South Asia, East Africa, and South America. The researchers found that allowing women to take leadership positions in community climate initiatives was instrumental in their long-term sustainability (Kratzer and Le Masson 2016). By contrast, initiatives in which women were sidelined failed to prosper over time (Kratzer and Le Masson 2016; Vazquez 2018).

The key is to bridge the gap between global ambitions for equality and on-the-ground realities. Women possess skills and ideas to solve climate problems (CSW-52 2008; UNDP 2013). It is important that both men and women are included in discourses to create effective and sustainable solutions. CDKN's work finds that discriminatory attitudes are principally to blame when women are left out—but it takes everybody to overcome them. Those with traditional decision-making power (who may not just be men but also the wealthier, more educated and privileged castes and classes) must be ready to make room at the table (e.g., Chaplin et al. 2019). At the same time, women must grow their confidence in their own knowledge and abilities and step up to be heard.

Countries can also make structural changes to level the playing field for women. While women tend to leave a smaller ecological footprint compared to men (Brough et al. 2016), owing to environmentally conscious behavior and attitudes, women still have comparatively lower ability to access the financial resources, technology, and information needed to adapt to the impacts of climate change. Greater access to finance and technology can increase their abilities to contribute to social resilience and to help societies adapt to the changing climate.

This chapter investigates the influence of women on environmental action in Pakistan, studying the main roles they play and showcasing stories of change that recognize the importance of engaging women on the environment. The women described here are "stepping up" to make themselves heard and to demonstrate environmental protection in practice. They are also accessing some of the vital forms of support mentioned—finance, technology, education, and information—to enable them to make a positive impact; and they are extending support to other women and girls to do the same.

Data and information were collected using an electronic survey ($n = 40$) and semi-structured interviews ($n = 15$) of respondents who are members of the Green Women Pakistan Forum.[3] The interviews brought five major roles to light, acknowledging both customary and modern perspectives on women. These roles were categorized as "environmental professionals," "home-managers," "policymakers," "green entrepreneurs," and "educators." Each is discussed at length in this chapter.

Roles of Women Within Environmental Action in Pakistan

Environmental Professionals

Female environmental professionals[4] have demonstrated capabilities as experts and innovators, making strides in environmental protection and climate action. These sustainability leaders work as sector specialists, environmental engineers, energy experts, health and safety operative assessors, economists, and scientists, to name a few. As such, these professionals need a fundamental understanding of the different ways that environmental systems interface with different sectors.

Pakistan has a family farming system, agriculture being an integral part of the economy. Women's role is central to food production, with about 72% of women associated with the agriculture sector (Farooq 2009). Global warming has affected food security, impacting crop seasons and productivity, while local environmental factors have impacted water and soil qualities (Ullah 2017). To support female farmers, particularly those that lack education and resources to cope with the changing climate and environment, Amna Jamil, an environmentalist, runs environmental clinics to provide female farmers with technical advice on agricultural decision-making. These clinics identify problems faced by female farming communities, assist in finding and implementing practical solutions, and develop capacities of farmers on resource usage in light of environmental and climate change.

Afia Malik, an environmental economist, has been working with corporations that are causing environmental harm to reduce and reverse such damage and help address the impact of environmental changes on their business value chains. Afia supports businesses in their transition to resource-efficient economic growth, and in identifying opportunities for such businesses to enter new and expanding "green" markets.

---

[3] Green Women Pakistan Forum was founded by Mehak Masood in 2014 as part of her green social entrepreneurship project for the Mosaic International Leadership Programme. The forum has an open membership policy for women working on the environment, studying environmental sciences and related subjects, or interested in environmental solutions. The Forum has accessible community platforms on Whatspp and Facebook, organizes dialogues with policymakers, and supports research on various environmental issues.

[4] Environmental professionals are experts who have a direct relationship with the environment; their professional responsibilities revolve around the natural environment, and they have the skills to solve real life environmental problems.

Her work with multinational corporations and industries has helped many businesses become resource-efficient, reducing their water and energy usage, increasing revenues, and becoming competitive in their markets. Using her expertise, some of her clients have also introduced products that are revolutionizing the green market in Pakistan, including a carpool mobile app, a smart water usage app for commercial laundries, and rainwater harvesting technologies and biogas digesters for household consumers.

Another influential leader has been Fatima Farooq. Working as an environmental lawyer and an environmental rights activist for two and a half decades, she now consults with top law firms in Pakistan, representing those most affected by environmental issues, primarily the vulnerable and poor. She has been instrumental in winning iconic environmental cases, setting historic penalties for environmental "crimes," and in establishing a national and provincial tribunal system for environmental protection. These women are just a few examples of the inspirational environmental professionals active in Pakistan.

Home-Managers

The customary roles of women in developing countries as caregivers to children, elders, and the sick, as home-managers, and as critical players in reproduction, makes them particularly susceptible to environmental threats, including pollution, climate change, and natural disasters (WHO 2017). These roles make women distinctly motivated to improve unsatisfactory conditions that affect their families and their own health (Wilson et al. 1996).

Women can be active defendants of the environment and equally responsible for the pressure they put on natural resources due to their contribution to population growth and pollution (Langer et al. 2015). This makes them ideal agents of change as day-to-day environmental managers to bring about societal behavioral change, motivate people to adopt greener, cleaner lifestyles, and instill environmentally friendly values in the next generation.

In Pakistan, housewives like Saima Malik, Riffat Jabeen, and Sadia Irshad have been at the forefront of environmental management and protection. In Multan, Saima Malik is a retired banker who works with secondary school children in her neighborhood in the evenings to raise their awareness of environmental problems and solutions and the role of each individual in society in environmental protection, using art techniques like painting, crafts, and drawing. Her neighborhood environmental campaign called "Green Me" is a festive summer affair and has been replicated in other communities across the city of Multan.

In Lahore, Riffat Jabeen, a retired English literature professor, has been successfully working with NGOs such as Hum Pakistan on urban forestry and waste management initiatives that have transformed into a participatory citizen force that identifies "trouble areas," raises funds from within the community, and uses youth volunteering to make practical changes to reverse environmental damages. She pioneered a rooftop garden at her own house, lives an almost zero carbon life, and aims to build a nationwide network of environmentally conscious housewives to promote greener lifestyles.

As a published researcher who has studied impacts of climate change on women in Pakistan, Sadia Irshad from Lahore has engaged with policymakers on regulations and approaches to supporting indigenous women's communities to fight climate change and has advocated pollution prevention with different youth groups. As a housewife, Sadia continues to work with indigenous women from suburban and rural areas all over the country to support poverty reduction and sustainable development activities.

According to the Women's Environment and Development Organization (WEDO 2007), women tend to share useful information with their communities that will help build their resilience and make them adaptable to environmental changes. Although communities benefit from women-led grassroots movements for environmental and climate action such as those led by housewives like Sadia, Saima, and Riffat, these initiatives will be more effective if women are better represented at decision-making levels.

Policymakers

There is irrefutable evidence that women are more environmentally conscious, and thus more likely to adopt environmentally friendly behavior such as recycling, buying eco-labeled products, and being more energy and fuel efficient, than men (Zelezny et al. 2002). They are more likely to showcase environmental identity by adopting environmentally protective behavior (Kashima et al. 2014), and more likely to opt for and support environmental policies (May et al. 2018).

A recent study from Pennsylvania State University shows that climate policy matters pertaining to science and business are attributed more to men while those concerning ethics and environmental justice are attributed to women (Swim et al. 2018). Another study suggests a better gender balance would result in more environmentally friendly policymaking (Cook et al. 2019). It is fair to conclude that gender not only has an impact on environmental attitudes, but also on decision and policymaking on the environment.

While women tend to support environmental and climate action, they are less likely to hold positions of power and to be included in decision-making groups. According to the national census 2017, women make up half of Pakistan's population, but only have 20% of (reserved) seats in the National Assembly (Latif et al. 2015). This system does not provide elected women with much power to influence decision-making or policy forming processes (Khattak and Hussain 2017).

Of the 40 female environmentalists included in this study, 90% thought that the national government was not doing enough to combat climate change and all of them agreed that the failure lies in the lack of female involvement in environmental policy making and governance. About 66% of respondents said their own work was hindered or affected by the limited presence of female decision makers on the environment.

Zakia Javaid is a seasoned deputy director at the Provincial Environmental Protection Department in Peshawar, Khyber Pakhtunkhwa, who has been at the helm of the development of provincial climate change policy and action plans for multilateral environmental agreements. As the only female working in this department, Zakia has often faced discrimination in the workplace and is not always permitted to travel to remote and rural areas for her work. Zakia shared: "Local representation of women in decision-making is limited due to the existence of age-old gender inequality challenges persistent in our cultures and customs. Women never get the political representation that is directly proportional to the female population in the country." Despite the challenges, Zakia has established interactive relationships with policy and civic communities to advance climate and environmental action. Her constant recommendation to the department's senior leadership is to increase women's participation as a pathway to enhancing effectiveness and the sustainability of climate change and environmental projects.

Green Entrepreneurs

Green entrepreneurs[5] have environmentally conscious and profitable business ventures that address environmental problems (Volery 2002), and provide creative, practical, and innovative solutions to address such concerns (Criscuolo and Menon 2015). These entrepreneurs innovate conventional approaches, structures, and methods (Parrish 2010), have established socioeconomic values (Kirkwood and Walton 2010) and tend to be key drivers of the green economy (Schaper 2010). Successful green entrepreneurial ventures focus on the creation of new services, products, or practices that address, reduce, or reverse environmental and social problems. Both innovation and creativity are fundamental for the success of such businesses.

Typically, creativity is associated with masculinity and eco-friendly behavior with femininity. Research shows a man's work is more likely to be considered creative and ingenious than a woman's (Proudfoot et al. 2015). However, studies have also shown that gender equality results in greater innovation. In fact, innovation is six times higher in organizations with a gender-balanced workforce (Shook and Sweet 2018). Since women have stronger environmental attitudes, are equally creative and innovative, and are more committed to growth, they are more likely to be engaged in green entrepreneurship than men (Braun 2010; Shmailan 2016).

[5] "Green," Eco, or Sustainable Entrepreneurship can mean businesses adopting sustainable practices, businesses developing environmentally friendly products or services, or addressing a specific environmental problem. In this instance, green entrepreneurs are those who have developed a product or service to address an environmental problem.

In Pakistan, Mehrunisa Malik and Farwa Tassaduq are two examples from among many other female entrepreneurs leading the green marketplace. Educated as environmentalists and experienced in climate and environmental action for more than a decade, both Mehrunisa and Farwa used their knowledge, creativity, and unorthodox approaches to launch their green businesses. Farwa uses biodegradable plastic and materials like used wood, cotton, and wool to develop eco-friendly toys for her company called One Earth Toys. Launched in Lahore in 2017, these non-toxic and environmentally friendly toys have been popular among the parenting community, which wants safe and intellectually stimulating toys for their children.

Mehrunisa, along with her partner Babar Naseem Khan, co-founded Saaf Suthra Sheher in 2016 to provide and mainstream responsible and ethical waste management solutions at the commercial and industrial level in Islamabad. They collect recyclable waste and ensure none of it goes to a landfill. The organization also helps set up social enterprises that spur urban regeneration and recycling and raises awareness through community outreach and capacity development. Both business ventures have been successful, but according to these green entrepreneurs, the success of green businesses largely depends on regulatory support and the environmental awareness of the public.

Educators

Education, both formal and informal, has a significant influence on the development of environmental awareness in both children and adults (Palmer et al. 1998). While regulations and governmental policies may influence citizens' environmental behavior, knowledge of environmental issues provides people with the basis for making lifestyle decisions that affect the environment. Understanding how the environment works and the impact of the actions of everyone on that environment is essential for the advancement of sustainable development (Khalili et al. 2015). Additionally, environmental educators, activists, researchers, and teachers raise awareness of environmental problems, issues, and possible solutions, particularly of actions that people can take to play their part in environmental protection (Gifford and Nilsson 2014).

A science teacher for secondary school learners, Hania Baloch has been teaching 12–15-year-olds about environmental problems in Karachi. She worked with the Department of Education on designing a curriculum that instills pro-environmental behavior in children. Hania shared: "We found it was important to evolve from just teaching children facts about environment and sustainability issues to educate them about the 'change' people need to make and see for the benefit of the environment." Hania and her fellow teachers formed a group in 2012 called Teach for Change (TfC) that uses a variety of non-formal settings, like after-school programs, camps, parks, zoos, outdoor centers, and so on, to host critical dialogues and discussions with children on selected topics pertaining to the environment and nature, and help them make good decisions in their lives.

Aysha Raja is a businesswoman and an environmental rights activist based in Lahore. She co-founded the group Citizens for Clean Air to address the air pollution problems plaguing Pakistan, to raise awareness among people of the issues and potential solutions, and to act as a pressure group for governmental action. The group organizes different public events, notably dialogues with government, and shares publicly accessible air quality monitoring data and information about the effects of air quality on health.

As a researcher, Mehak Masood has investigated policies, tools, and procedures that support the path of transition to sustainable development in Pakistan. Her research to date has helped introduce the concept of circular economy (redirecting waste from disposal to product manufacturing in order to reduce waste production) in industrial estates in Punjab. This has led to streaming waste from one industry, as raw material, to another, and has nearly eliminated waste disposal on industrial sites. She also developed a roadmap for public–private partnerships for municipal waste management, which garnered private sector interest in waste management. Presently, she is exploring community engagement models for national and provincial environmental departments to better integrate communities in environmental governance. Mehak considers research to be a structural element for environmental awareness, sharing information with a wider audience, and "using research as a platform for youth to come and connect to policymakers."

Conclusion

At the heart of the gendered debate on climate action is the fact that women have experienced the impacts of the degraded environment and changing climate for years and have since been leading environmental protection efforts as managers, rehabilitators, and innovators of the natural environment. Their knowledge, expertise, and unique experiences in fighting climate change represent an untapped asset in coping with environmental issues. Moreover, there is potential to inculcate their pro-environmental values within the future generations and their communities.

While women are valuable agents of change, they are not often recognized or supported in their endeavors. Yet, women are playing an integral role in contributing solutions to environmental and climate challenges as environmental professionals, policymakers, green entrepreneurs, and educators. As home-managers, women have an even more significant role to play in shaping the environmental behavior of future generations, and in influencing family and community values and lifestyles.

These five roles place women at the heart of climate action, providing space for women to deliver immense benefits to communities and societies. If better integrated into policymaking and environmental planning—and with access to relevant education, trainings, financial resources, and technologies—women can further enhance, expand, and develop their role as protectors of the environment.

Acknowledgements This research study is part of the Green Women Pakistan Forum's program on "Mobilizing Environmental Action in Pakistan." Our gratitude to The Green Women Pakistan Forum for continuous support and encouragement.

References

Aguilar Revelo, Lorena. 2009. "Training Manual on Gender and Climate Change." The International Union for Conservation of Nature website.

Braun, P. 2010. "Going Green: Women Entrepreneurs and the Environment." *International Journal of Gender and Entrepreneurshi*p 2: 245–259. https://doi.org/10.1108/17566261011079233.

Brough, A. R., J. E. Wilkie, J. Ma, M. S. Isaac, and D. Gal. 2016. "Is Eco-Friendly Unmanly? The Green-Feminine Stereotype and Its Effect on Sustainable Consumption." *Journal of Consumer Research* 43, no. 4: 567–582. https://doi.org/10.1093/jcr/ucw044.

Chaplin, D., J. Twigg, and E. Lovell. 2019. *Intersectional Approaches to Vulnerability and Resilience-Building*. London: ODI and BRACED.

Cook, N. J., T. Grillos, and K. P. Andersson. 2019. "Gender Quotas Increase the Equality and Effectiveness of Climate Policy Interventions." *Nature Climate Change* 9, no. 4: 330–334. https://doi.org/10.1038/s41558-019-0438-4.

Criscuolo, C., and C. Menon. 2015. "Environmental Policies and Risk Finance in The Green Sector: Cross-Country Evidence." *Energy Policy* 83: 38–56. https://doi.org/10.1016/j.enpol.2015.03.023.

CSW-52 (Commission on the Status of Women—52nd Session). 2008. "Gender Perspectives on Climate Change." *Issues Paper for Interactive Expert Panel: Emerging Issues, Trends and New Approaches to Issues Affecting the Situation of Women or Equality between Women and Men*. Accessed June 6, 2019.

Farooq, O. 2009. *Economic Survey of Pakistan*. Government of Pakistan.

Gerretsen, Isabelle. 2018. "Half of Climate Action Plans Neglect People in Danger, Say Researchers." *Thomson Reuters* website. July 12, 2018. https://www.reuters.com/article/us-climatechange-adaptation-development/half-of-climate-action-plans-neglect-people-in-danger-say-researchers-idUSKBN1K12PF.

Gifford, R., and A. Nilsson. 2014. "Personal and Social Factors that Influence Pro-Environmental Concern and Behaviour: A Review." *International Journal of Psychology* 49, no. 3: 141–157. https://doi.org/10.1002/ijop.12034.

IPCC (Intergovernmental Panel on Climate Change). 2019. *Special Report on Climate Change and Land*. Geneva: IPCC. Chapter 7: Cross Cutting Box 11 (Gender in inclusive approaches).

Kashima, Y., A. Paladino, and E. A. Margetts. 2014. "Environmentalist Identity and Environmental Striving." *Journal of Environmental Psychology* 38: 64-75. https://doi.org/10.1016/j.jenvp.2013.12.014.

Khalili, N. R., S. Duecker, W. Ashton, and F. Chavez. 2015. "From Cleaner Production to Sustainable Development: The Role of Academia." *Journal of Cleaner Production* 96: 30–43. https://doi.org/10.1016/j.jclepro.2014.01.099.

Khattak, S. S., and A. Hussain. 2017. "Women Representation in Pakistani Legislatures." *South Asian Survey* 20, no. 2: 191–205. https://doi.org/10.1177/0971523116679785.

Kirkwood, J., and S. Walton. 2010. "How Ecopreneurs' Green Values Affect their International Engagement in Supply Chain Management." *Journal of International Entrepreneurship* 8: 200-217. https://doi.org/10.1007/s10843-010-0056-8.

Kratzer, S., and V. Le Masson. 2016. *10 Things to Know: Gender Equality and Achieving Climate Goals*. London: ODI and Climate and Development Knowledge Network.

Langer, A., A. Meleis, F. M. Knaul, R. Atun, M. Aran, H. Arreola-Ornelas, Z. A. Bhutta, A. Binagwaho, R. Bonita, J. M. Caglia, M. Claeson, J. Davies, F. A. Donnay, J. M. Gausman, C. Glickman, A. D. Kearns, T. Kendall, R. Lozano, N. Seboni, G. Sen, S. Sindhu, M. Temin, and

J. Frenk. 2015. "Women and Health: The Key for Sustainable Development." *The Lancet* 368: 1165–1210. https://doi.org/10.1016/S0140-6736(15)60497-4.

Latif, A., A. Usman, J. R. Kataria, and M. Abdullah. 2015. "Female Political Participation in South Asia: A Case Study of Pakistan." *South Asian Studies* 30, no. 2: 201–203.

Manuel, M, F., Grandi, S., Manea, Kirbyshire, A., and E. Lovell. 2018. *Leave No One Behind Index*. London: ODI.

May, A. M., M. G. Mcgarvey, and D. Kucera. 2018. "Gender and European Economic Policy: A Survey of the Views of European Economists on Contemporary Economic Policy." *Kyklos* 71, no. 1: 162–183. https://doi.org/10.1111/kykl.12166.

Palmer, J., J. Suggate, B. Bajd, P. Hart, R. Ho, J. Ofwono-Owecho, M. Peries, I. Robottom, E. Tsaliki, and C. V. Staden. 1998. "An Overview of Significant Influences and Formative Experiences on the Development of 115 Adults' Environmental Awareness in Nine Countries." *Environmental Education Research* 4: 445–464.

Parrish, B. D. 2010. "Sustainability-Driven Entrepreneurship: Principles of Organization Design." *Journal of Business Venturing* 25: 510–523.

Peterson, K. 2007. "Reaching Out to Women When Disaster Strikes." *Soroptimist White Paper*. http://www.soroptimist.org/.

Proudfoot, D., A. C. Kay, and C. Z. Koval. 2015. "Gender and Perceived Creativity." Master's thesis, Duke University. https://doi.org/10.1177/0956797615598739.

Schaper, M. (ed.). 2010. *Making Entrepreneurs: Developing Sustainable Entrepreneurship*. Gower: Farnham.

Shmailan, A. B. 2016. "Compare the Characteristics of Male and Female Entrepreneurs as Explorative Study." *Journal of Entrepreneurship and Organization Management* 5: 203.

Shook, E., and J. Sweet. 2018. "When She Rises, We All Rise." Accenture website.

Swim, J. K., T. K. Vescio, J. L. Dahl, and S. J. Zawadzki. 2018. "Gendered Discourse about Climate Change Policies." *Global Environmental Change* 48: 216–225. https://doi.org/10.1016/j.gloenvcha.2017.12.005.

Ullah, S. 2017. "Climate Change Impact on Agriculture of Pakistan—A Leading Agent to Food Security." *International Journal of Environmental Sciences and Natural Resources* 6, no. 3.

UN/ADPC. 2010. "Disaster Proofing the Millennium Development Goals (MDGs)." By the UN Millennium Campaign and the Asian Disaster Preparedness Center. Accessed June 5, 2019. http://www.adpc.net/v2007/downloads/2010/oct/mdgproofing.pdf.

UNDP. 2013. *Gender and Disaster Risk Reduction*. United Nations Development Programme website. https://www.undp.org/content/dam/undp/library/gender/Gender%20and%20Environment/PB3-AP-Gender-and-disaster-risk-reduction.pdf.

UNFCCC. 2015. "Paris Agreement on Climate Change." United Nations Framework Convention on Climate Change website. Accessed October 1, 2019.https://unfccc.int/process-and-meetings/the-paris-agreement/the-paris-agreement.

UNFCCC. n.d. "The Enhanced Lima Work Programme on Gender." United Nations Framework Convention on Climate Change website. Accessed October 1, 2019. https://unfccc.int/topics/gender/workstreams/the-lima-work-programme-on-gender.

Vazquez, A. 2018. *Synthesis Report: Contributions of the Climate-Resilient Cities in Latin America Initiative*. Quito: Fundacion Futuro Latinoamericano and Climate and Development Knowledge Network.

Volery, T. 2002. "Ecopreneurship: Rationale, Current Issues and Futures Challenges." In *Radical Change in the World—will SMEs Soar or Crash?* edited by U. Fueglistaller, T. Volery, and W. Weber, 541–553. St. Gallen: KMU-HSG.

WEDO. 2007. *Changing the Climate: Why Women's Perspectives Matte*r. New York: Women's Environment and Development Organization.

WHO. 2017. *Don't Pollute My Future! The Impact of the Environment on Children's Health*. Geneva: World Health Organization.

Wilson, M., M. Daly, S. Gordon, and A. Pratt. 1996. "Sex Differences in Valuations of the Environment?" *Population and Environment* 18: 143–159.

Zelezny, L. C., P. Chua, and C. Aldrich. 2002. "New Ways of Thinking about Environmentalism: Elaborating on Gender Differences in Environmentalism." *Journal of Social Issues* 56, no, 3: 443–457. https://doi.org/10.1111/0022-4537.00177.

Chapter 16
Development Challenges in Papua and West Papua

Alex Rumaseb

Highlights

- Development challenges in West Papua are complex; in particular, separation conflicts and power struggles constrain the development of local civil society.
- The author, a former policymaker and now a sustainable development observer (i.e., working outside the government), experienced a degree of transformational learning, emphasizing that giving voice and representation in planning to diverse indigenous views and ensuring various government-enabled conditions are central to healthy civic engagement.
- West Papuan stakeholders are supported by generous Special Autonomy and associated funds, but too much money can also co-opt stakeholders, including civil society.

Introduction

The Indonesian provinces of Papua and West Papua—which together comprise the western half of the island of New Guinea and for the purposes of this paper will be collectively termed *West Papua*—are rich in both indigenous cultures and natural resources. About 250 ethnic groups live in the *West Papua* provinces (KPE-IJ 1993), with some of the languages spoken by fewer than 1,000 people.

The tectonic history of the island of New Guinea has formed high biodiversity environments, including pristine mountain ranges, prominent lakes, swamps, mangroves, and coral reefs (Marshall and Beehler 2007). *West Papua* has the most extensive and least disturbed tropical moist forest cover in Indonesia, as well as Indonesia's largest mangrove ecosystem, at Bintuni Bay. The northwestern part of

Thanks are due to Agus Sumule for providing constructive criticism of the manuscript.

A. Rumaseb (✉)
Planning Agency, Jayapura, Indonesia

M. Indrawan et al. (eds.), *Civic Engagement in Asia*,
https://doi.org/10.1007/978-981-16-9384-7_16

the island contains the world's greatest diversity of coral reef fish, with more than 1,650 species, many of which can be found nowhere else in the world (Pattiselanno and Arobaya 2014). The wide range of ecological and geographical diversity allowed each ethnic group to have distinct territories, cultures, and languages. The unique ecological, linguistic, and ethnographic characteristics in turn have given rise to a constellation of unique social, political, and economic conditions.

Unfortunately, *West Papua* continues to face serious development challenges, especially in terms of socioeconomic and ecological sustainability. Today Papua and West Papua are among the poorest of all Indonesian provinces. The Statistics Office of Papua province recorded in September 2019, for example, that the rate of rural poverty was 35.36%,[1] *versus* 4.53% for urban poverty.[2] Further, data in September 2019 showed that poverty levels in both Papua and West Papua provinces (respectively 26.55[3] and 21.51%[4]) remained more than twice the national average, which is 9.22%.[5]

Indigenous *West Papuans* mainly live in small villages and at forest margins. The national government implemented large-scale infrastructure development (Sloan et al. 2019) and supported the conversion of natural ecosystems into large-scale oil palm plantations; all this was done without adequate provisions for consulting with local governments and stakeholders, including indigenous people (Indrawan et al. 2017).

This chapter reflects on the development challenges of *West Papua* from the author's evolving points of view, i.e., from bureaucracy to civil society. First, the Papuan conflict that continuously impacts development prospects will be briefly introduced. Development challenges in the public sectors of health, education, and the environment will then be highlighted. Finally, a way forward that is inclusive of cultural perspectives will be considered.

Conflict and Struggle for the Independence of *West Papua*

The *West Papuan* provinces are the location of the longest-running violent struggle for secession in Indonesia and the Pacific. The chronic conflict is rooted in the differing perspectives between the central government of Indonesia and indigenous Papuans

---

[1] Additionally, based on my own observations, I would estimate that roughly 70% of the rural poor in these provinces were indigenous *West Papuans*.

[2] https://papua.bps.go.id/pressrelease/2020/01/15/476/profil-kemiskinan-di-provinsi-papua-september-2019.html [official press release, "Poverty Profile for Papua Province as per Sept 2019"] (in Bahasa Indonesia).

[3] www.papua.bps.go.id.

[4] www.papuabarat.bps.go.id.

[5] Government statistics for Indonesia and its provinces can be studied at the Bureau of Statistics of Indonesia website, which offers some page navigation and statistical information in English, more in Bahasa Indonesia: https://www.bps.go.id.

regarding the territory's integration into the nation of Indonesia following its independence. This has been exacerbated by decades of political violence, human rights violations, development failures, and the marginalization of indigenous Papuans (Widjojo 2010; MacLeod 2014; Anderson 2015; Korwa 2016; Pamungkas 2017; Viartasiwi 2018).

Violence associated with *West Papua*'s struggle for independence has at times reached disastrous proportions. Even during recent years (2009–2014), 166 cases of violence were documented in the Papua Regional Police Threat Perception Report (Syailendra 2016). Still more recently, the independence movement has transformed into a partially nonviolent resistance that utilizes mass media and social campaigns within and beyond Indonesia (Pamungkas 2017; Ruhyanto 2018). The government of Indonesia is also adapting by enhancing the roles and capacities of the local police in counter-insurgence, expanding conventional intel-gathering and law enforcement roles to include social empowerment and opinion-making approaches (Syailendra 2016). Nevertheless, violence has continued from both sides (Smith 2018).

Considering the ongoing conflicts throughout the history of the integration of *West Papua* into Indonesian territory, two issues need to be clarified. First, moves for *West Papuan* secession have never been truly united. Differentiated factions proliferated from the ranks of local *West Papuan* actors and stakeholders, including civil society organizations, government bodies, and communities of activists overseas; these many groups have never acted as a single centralized movement (Timmer 2005). Second, in the highlands of Papua, the role of the State has been negligible. Instead of dominance, there seems to be an absence, unlike other highlands of Southeast Asia (Anderson 2015).

Development Challenges

Significant changes have come quite abruptly to the region. No less than 50 years ago, indigenous Papuans followed subsistence-based livelihoods, living off nature's bounty. This is especially the case in interior areas, where indigenous communities have tended to adhere to traditional land tenure systems and where communal living, reliance upon oral traditions, bark clothing, tribalism and tribal wars, and a barter economy (instead of a cash-based economy) have historically been central aspects of community life.

The government of Indonesia's mass transmigration program in the 1970s brought large numbers of ethnically Javanese people to Papua (Pouwer 1999; Upton 2019)[6] and was followed by Special Autonomy for *West Papua* a few decades later, beginning

[6] The first wave of migration during the 1970s was predominantly comprised of Javanese, but ensuing inbound migrations came from throughout the archipelago. According to a 2000 census, non-Javanese comprised 62% of the settler population in Papua (McGibbon 2006).

in 2001.[7] Both of these Central Government policies imposed serious and sudden systematic political and cultural changes. Before the advent of Special Autonomy, state development policies alienated indigenous people, who were used to communal, not private, rights. Government policies often favored heavily capitalized resource extraction in the name of "development" or "economic growth," and this gradually marginalized the indigenous people of *West Papua*.

Changes in demographic patterns, due to transmigration, further marginalized indigenous people (Ananta et al. 2018). The total population of *West Papuan* provinces is now close to 5 million people, which is more than three times the population at the beginning of transmigration. In 1971, indigenous Papuans represented 96% of the population of *West Papua*, but they have since suffered displacement from their traditional lands, especially in more accessible coastal areas, where the indigenous population is now only 30% of the total.

After decades of alienation from development policies, the abundant, if temporary, finances that the Special Autonomy system disbursed to the region—which were not accompanied by adequate capacity building in development planning or implementation[8]—contributed to the unwanted side effect of increased consumptive behavior on the part of the tribal chiefs and villagers. At the provincial level, it also led to the rise of gross inefficiency in development spending. At the same time, neither the resources from the national government nor from international donors could be effectively channeled, creating a common refrain among development partners that "Papua is not lacking in financial resources, but in absorptive capacities."

Even after the advent of Special Autonomy in Papua, development efforts implemented without proper understanding of the local context have at times fallen flat with respect to expected results, or worse, reinforced existing inequalities. For example,

---

[7] This included a string of national-level policies:

(i) Law No. 21/2001 regarding Special Autonomy for Papua;

(ii) Presidential Decree No. 1/2003 for the acceleration of the implementation of Law No. 45/1999, which mandated the creation of two new provinces (West Irian Jaya and Central Irian Jaya), three new districts, and one new city. However, in this provincial blossoming, originally from Papua, only West Irian Jaya followed the legal protocols. From then on, de facto and de jure, the western part of Indonesian New Guinea had two provinces, namely Papua and West Irian Jaya (which in 2007 was renamed West Papua Province);

(iii) Government Regulation No. 24/2007 regarding the Name Change of West Irian Jaya Province to West Papua Province.

(iv) Government Regulation in lieu of the Law No. 1/2008 regarding amendment to Law 21/2001 regarding Special Autonomy of Papua.

[8] The Special Autonomy Laws granted the provincial government of Papua (and following the provincial split, the West Papua provincial government as well) greater decision-making powers in every sector except international affairs, defense, monetary and fiscal policy, religion, and justice. The Law also mandated that a considerably higher share of the revenue originating in *West Papua* stay in *West Papua*—stipulating that until 2026 both the provinces of Papua and West Papua would receive 80% of the revenues from their natural resources sectors (including forestry, fisheries, and mining), and 70% of the initial revenues from oil and gas, 50% thereafter. Furthermore, on top of the financial transfers from the central government that all provinces have been receiving, until 2021 Papua has received additional special autonomy funds amounting to 2% of the total national General Purpose Fund (Resosudarmo et al. 2014).

due to a lack of consultation, even road infrastructure, which in theory should help boost the local economy, has exposed certain problems. For instance, many indigenous Papuans continue to rely on walking as their primary form of transportation, because if they use vehicles to access markets (for example to sell raw bananas), the net profit barely covers the cost of the trip. The market economy itself is hardly known by most of the indigenous peoples, who have relied on hunting and gathering all their lives and do not have extensive experience in producing value-added items for the marketplace. With the influx of migrants from Java and other parts of Indonesia came free trade, capitalism, and the cyberculture. All these potentially pose positive and negative impacts. As such, the economic opportunities provided by new road infrastructure are more likely to place newcomers (who for example might sell fried bananas in the market) at an advantage instead of indigenous peoples. Newcomers may cement their centrality in this new road-oriented economy by developing permanent buildings and an ever-growing number of roadside kiosks. Pointing to the contradictions in the promise of "development," Timotius Murib, a Papuan leader exclaimed: "We don't need development, because with development we lose control of our land" (BBC 2018).

Since the rapid and poorly planned transition to a cash economy precipitated by Special Autonomy prompted materialism and consumerism among indigenous communities, the spirit of communalism and appreciation of nature has been replaced by the hunt for development money. Today, many indigenous peoples are just as likely to spend days at government offices with (often poorly designed) proposals seeking handouts from development projects, as they are to spend time pursuing their traditional livelihoods.

Rapid urbanization has also led to a decline in traditional farming practices. Many younger people have migrated to the coastal cities seeking work, but due to inadequate skill sets, have joined the ranks of the urban poor. Formerly, the city of Wamena and its surrounding areas in the central cordillera produced yams, taro, and vegetables to feed Jayapura and other large cities of Papua province. Now mainly only older people remain, with the majority of gardens having been left untended, threatening in this way *West Papua's* own food security. These days swelling urban populations increasingly rely on food imported from other islands.

In-migration has also had political implications. Many non-Papuan migrants are now taking over positions of power, including in the legislature. For example, of the 45 newly elected (2019) members of the Legislature of West Papua province (DPRPB), only 17 were indigenous Papuans. An increasing number of migrants are also taking civil service positions in municipalities, districts, and villages. This reflects today's population of *West Papua*, which, especially in the coastal areas, is dominated by migrants and newcomers.

As the indigenous people were marginalized by these developments, they began to sell their land, often to cover immediate livelihood needs and the exorbitant costs of their children's schooling. Their traditional subsistence-based cultures have tended not to include strong concepts of entrepreneurship and long-term investment, which would be needed were they to successfully compete with new migrants in this new and alien (from an indigenous perspective) economic era. Consumerism and alcoholism further contribute to the selling of indigenous land. In some instances, large-scale plantation agribusiness actors have coerced indigenous people into accepting gifts and alcohol in exchange for access to their lands, and previously natural ecosystems have been converted into agriculture (Mongabay and Gecko Project 2019).

Health and Education Issues

Indonesia's Central Bureau of Statistics reported the 2010–2015 Human Development Index (HDI)[9] for the provinces of West Papua and Papua as the lowest of all Indonesian provinces. West Papua's HDI score was 61.73 and Papua's 57.25, both well below the Indonesian average of 69.55. Access to decent education and healthcare are particularly challenging. Since Papua was given Special Autonomy status in 2001, the central government has provided more than $3 billion USD in additional funding, with plans to provide more before 2025, when the special status expires. Illiteracy remains a major problem in *West Papua*, with an overall illiteracy rate of close to 30% and up to 90% in the highland districts (Grainger 2014). Some teachers do not demonstrate adequate literacy and numeracy skills, and many children are unable to read or write despite having finished elementary school (Grainger 2014). These problems stem from multiple sources, including a shortage of teachers and school facilities, lack of community support (children are often expected to help parents in the fields, rather than go to school), corruption by sector officials, and geographic impediments. For example, it is not uncommon for elementary school students to have to walk 40 minutes to attend school (RCA 2015).

West Papua and Papua provinces serve as prime examples of the inequities of access to health care and knowledge in Indonesia. Even as recently as 2010, West Papua and Papua provinces occupied first and third place nationally in terms of maternal mortality. The maternal mortality rate in Papua province of 573/100,000 live births in 2010 (BPS 2013) was more than twice the national average (Nababan et al. 2018). Similarly, the infant mortality rate of 115/1000 live births (BPS 2013) is well above average.[10] The immunization coverage rate of 55.8% in Papua province is well below the national rate of 89.5% (DPR RI 2014). Papua province also has the highest HIV prevalence in Indonesia (Yunifiar et al. 2017), while HIV prevalence across both provinces (Papua and West Papua) reached 2.3%, with 2.9% of indigenous Papuans carrying the virus in 2013[11] (Munro 2015). Several high-end hospitals have recently

---

[9] The HDI provides a measure of the average achievements of citizens in terms of human development. The variables include general health and life expectancy, education, and living standards.

[10] The fluctuation, however, may be considerable: in less than 5 years the maternal mortality rate declined to 380/100,000 live births and the infant mortality rate declined to 13/1000 live births (Republika 2017).

[11] These are results of the 2013 Integrated Bio-Behavioural Survey (IBBS), which explored HIV's multiple indicators throughout the provinces of Papua and West Papua. Matters related to prevalence, knowledge of HIV, sexual behavior, and condom use were presented by Dr. Arnoldus Tiniap from

opened in Jayapura and larger cities. Yet, most indigenous *West Papuans* live in remote villages far away from these cities. It would be more beneficial to have a strengthened "posyandu" (an integrated service system) that reaches the remote areas where many indigenous Papuans live.

Navigating Through High Carbon Development

From the onset of Special Autonomy, *West Papua* has embarked on a risky path in which natural resources (including forested lands) are not treated as long-term investments, but rather as raw materials for extraction. Indeed, extraction accounts for 70% of the *West Papua* economy, with only 30% of economic activity occurring in more tertiary sectors such as services, trade, and infrastructure (Rumaseb, unpublished data). Incoming funds from the central government and international donors have not been effectively used at the district and village levels; furthermore, at the household level, borrowing often exceeds savings. Returns on investment have tended to flow to other islands, as the more influential economic investors originate from outside *West Papua*. All these realities have given rise to the notion of the *West Papuan* paradox, that indigenous *West Papuans* live in absolute poverty amidst rich natural resources.

At one stage, the development discourse of *West Papua* steered toward sustainability. In 2010, Papua province, with the assistance of the UK government as an international development partner, formulated its Vision 2100 plan, which provided a long-term vision of sustainable development based on a more efficient economy that aimed at eliminating dependence on natural resources exploitation and large-scale forest conversion. The vision presented a development model that facilitated improved and more equitable welfare, empowered indigenous people through proper participation in planning, and encouraged local control of energy security and local food production through sustainable natural resource management. This vision was supplemented by the necessary spatial planning and the passage of a provincial regulation in 2013[12] that mandated the maintenance of 90% of Papua province's forest cover. Yet, this vision has not been sustained after its champion, Barnabas Suebu, the former governor of Papua province, left office. However, recent high-profile sustainability events involving both Papua and West Papua provinces brought the matter back into discussion (Indrawan et al. 2019; Cámara-Leret et al. 2019).

West Papuan health agency at a November 2014 workshop, "Developing an HIV Prevention and Control Strategy for Papuans in Tanah Papua".

[12] Peraturan Daerah No. 23 Tahun 2013 tentang Rencana Tata Ruang Wilayah (RTRW) Provinsi Papua [Regional regulation no 23 of the Year 2013 regarding Spatial Plans for the Province of Papua] (in Bahasa Indonesia).

Discussion: The Way Ahead

Multiple challenges are standing in the way of sustainable development for *West Papuan* provinces. Regarding the vicious cycle of conflict and violence between the state and insurgent forces, there should be increased understanding that rebels will consider confrontational responses from the state as non-accommodating. Convening human rights courts and revealing the facts of past incidents may facilitate reconciliation (ELSAM 2014). Continuing insurgent efforts are a challenge that reflects the indigenous need to be properly involved in the development of Papua, especially human development. The social and political stigmatization of indigenous Papuans (using condescending terms such as "backwards" for example) also continues to impede progress in *West Papua*.

It has been said that *West Papua* is an anthropological dilemma that was treated with political approaches.[13] There is also the old paradigm, that *West Papuans* retained the cultural view of cargoism, a widespread belief in the Melanesian cultures outlining how indigenous peoples can obtain cargo (material possessions) similar to that held by Europeans through the use of rituals. The belief entailed the expectation that a messianic figure would come and deliver the valuables (McDowell 1994). This widespread belief's misalignment with conventional ideas of "development" presents an additional challenge.

It may not be fair to place all the blame for development impediments on the national government. *West Papuan* provinces have their own challenges in terms of leadership and representation. For instance, the *West Papuan* elite used Special Autonomy to reciprocally co-opt the state (Anderson 2015). In other words, disorder provided opportunities for elite *West Papuans* who knew how to intervene in the system for their personal benefit, and sometimes to the detriment of the broader population (Timmer 2005).

In many instances, *West Papuan* development is not turning to, or connecting with, sustainable development principles and inclusive participation. The customary traditional structure that was centered around a "big man" and a "chief" (White 2007; McLeod 2008)[14] has been corrupted in recent years, especially when the customary leaders, who have the trust of the communities, suddenly gained access to development funds without the necessary knowledge and skills in contemporary management.

---

[13] From repeated comments by indigenous participants during the training series "Sustainable development: towards Green and Sustainable Papua" (August 2009–July 2010, 9 workshops), which was a cooperation between the Papua Planning Agency and the UK Department for International Development, with unpublished summary notes written by Nusya Kuswatin in 2010.

[14] The two differ markedly. "Big Man" is characterized by personal power: status is earned through the demonstration of skills (magic, oratory, bravery); status is earned and maintained via generosity in the distribution of wealth; influence is exerted over fluctuating factions. In contrast, with the Chief, power resides in the position, not the person: status is inherited (not achieved); authority is exerted over permanent groups; authority to call upon the support of others is without inducement (McLeod 2008).

West Papuan civil society is also challenged, as many NGOs and religious organizations have become lax and compromised, if not corrupted, by the abundant Special Autonomy funds. Civil society has yet to be strengthened sufficiently to participate effectively in development partnerships to formulate, monitor, and evaluate development plans, including spatial plans. More recently, there are NGOs who have come to *West Papua* with a set of values, ideas, and assumptions that did not engage with Papuans' diverse cultural perspectives. For instance, the author observed that the NGO supported programs were often not sensitized to the heterogeneity of perspectives, e.g., the cultures of coastal inhabitants would markedly differ from the highland tribes.

On the other side, the national government has yet to listen to the democratic needs of the indigenous people. Whereas a centrally driven economic approach to development from the national government has resulted in better facilities and infrastructure, local perspectives of how local development and resources should be governed, and affirmative action to ensure more egalitarian benefit distribution, have yet to be mainstreamed. Unless these issues are resolved through serious consultation, insurgency will always result.

West Papuan provinces do not need to meet an imposed uniformity of development that may be acceptable in Indonesia's other provinces, but which is inappropriate for the local context. Development planning challenges, including inaccurate spatial planning data, need to be resolved. To this end, the nationwide One Map Policy that is nearing completion should be fully supported. Furthermore, although Papua province created a blueprint for sustainable development as far back as 2009, it has not been consistently implemented by the changing political regimes in Papua and West Papua provinces. Nonetheless, more recently there are promises of joint measures for sustainability between Papua and West Papua provinces (Indrawan et al. 2019; Cámara-Leret et al. 2019).

At the moment, there seem to be more challenges than solutions. It is important that any solution must begin with building and appreciating the perspectives of the *West Papuan* indigenous peoples as much as the national and provincial governments. Mutual understanding has yet to be built but remains the singular prerequisite for building West Papua from within.

References

Ananta, A., D.R.W.W. Utami, and N.B. Handayani. 2018. "Statistics on Ethnic Diversity in the Land of Papua, Indonesia." *Asia & the Pacific Policy Studies* 3 (3): 458–474. https://doi.org/10.1002/app5.143.

Anderson, B. 2015. "Papua's Insecurity: State Failure in the Indonesian Periphery." *Policy Studies* 73. Washington, DC: East-West Center.

BBC News. 2018. "Indonesia Attack: Gunmen Kill 24 Construction Workers in Papua." December 5. https://www.bbc.com/news/world-asia-46446719.

BPS (Bureau of Statistics Indonesia, National Population and Family Planning Board/ BKKBN, Ministry of Health Indonesia and ICF International). 2013. *Indonesia Demographic and Health*

Survey 2012. Jakarta: Bureau of Statistics Indonesia, National Population and Family Planning Board (BKKBN), Ministry of Health Indonesia and ICF International.

Cámara-Leret, R., A. Schuiteman, T. Utteridge, G. Bramley, R. Deverell, L.A. Fisher, J. McLeod, L. Ha, P. Roehrdanz, T.G. Laman, E. Scholes, Y. de Fretes, and C. Heatubun. 2019. "The Manokwari Declaration: Challenges Ahead in Conserving 70% of Tanah Papua's Forests." *Forest and Society* 3: 148–151.

DPR RI. 2014. Unpublished Report. Working Visit of Commission IX of the House of Representatives of the Republic of Indonesia on 7–10 December 2014.

ELSAM. 2014. "Mendorong pembentukan kembali UU Komisi Kebenaran dan Rekonsiliasi" [Revitalizing Measures for the (Legal) Enactment of the National Commission for Truth and Reconciliation] (in Bahasa Indonesia). *Policy Brief*. Accessed July 23, 2019. https://referensi.elsam.or.id/wp-content/uploads/2014/09/5.-Mendorong-pembentukan-kembali-komisi-kebenaran-dan-rekonsiliasi.pdf.

Grainger, S. 2014. "Remote Schools: Inspiring Papua's Children to Go the Extra Mile for Education." UNICEF Indonesia Blogpost, April 14. http://unicefindonesia.blogspot.com/2014/04/remote-schools-inspiring-papuas.html.

Indrawan, M., J. Caldecott, and Ermayanti. 2017. "Mitigating Tensions over Land Conversion in Papua, Indonesia." *Journal of Asia Pacific Policy Studies* 4 (1): 147–157. https://doi.org/10.1002/app5.157.

Indrawan, M., A. Sumule, A. Wijaya, N. Kapisa, F. Wanggai, M. Ahmad, B.V. Mambai, and C.D. Heatubun. 2019. "A Time for Locally Driven Development in Papua and West Papua." *Development in Practice*. https://doi.org/10.1080/09614524.2019.1609907.

Korwa, J.R.V. 2016. "Earned Sovereignty: The Road to Addressing Prolonged Conflict in West Papua—Indonesia." Master's Thesis, Flinders University.

KPE-IJ (Kelompok Peneliti Etnografi Irian Jaya). 1993. "Etnografi Irian Jaya: panduan sosial budaya" [Ethnography of Irian Jaya: Guide to the Local Social and Cultural Dimension] (in Bahasa Indonesia). Jayapura: Kelompok Peneliti Etnografi Irian Jaya.

McLeod, A. 2008. "Leadership Models in the Pacific." *State, Society and Governance in Melanesia*. Discussion Paper 2008/6 of the Research School of Pacific and Asian Studies, Australian National University. https://openresearch-repository.anu.edu.au/bitstream/1885/10082/1/McLeod_LeadershipModels2008.pdf.

MacLeod, J. 2014. "From the Mountains and Jungles to the Villages and Streets: Transitions From Violent to Nonviolent Resistance in West Papua." In *Civil Resistance and Conflict Transformation: Transitions from Armed to Nonviolent Struggle*, edited by D. Véronique. New York: Routledge.

Marshall, A. J., and B. M. Beehler, eds. 2007. *The Ecology of Papua*. Singapore: Periplus.

McDowell, M. 1994. "The Contextualization of Cargo Cult Beliefs and the Christian Message in Irian Jaya, Indonesia." PhD Dissertation, Oxford University. https://pdfs.semanticscholar.org/d63f/4359493c9a0cbc35ec8a67bc6570cefea9b8.pdf.

McGibbon, R. 2006. "Pitfalls of Papua: Understanding the Conflict and Its Place in Australia-Indonesia Relations." *Lowy Institute Paper 13*.

Mongabay and Gecko Project. 2019. "How Land Grabbers Co-Opt Indigenous Traditions in Papua." Q&A with Anthropologist Sophie Chao. https://news.mongabay.com/2019/03/how-land-grabbers-co-opt-indigenous-ritual-traditions-in-papua-qa-with-anthropologist-sophie-chao/.

Munro, J. 2015. "The 2013 Survey on HIV in Tanah Papua: Gender, Ethnicity and 'Casual Partners.'" In Brief No 11. Australian National University.

Nababan, H.Y., M. Hasan, T. Matthias, R. Dhital, A. Rahman, and I. Anwar. 2018. "Trends and Inequities in Maternal Health Care Services in Indonesia, 1986–2012." *International Journal of Women's Health* 10: 11–24.

Pamungkas, C. 2017. "The Campaign of Papua Peace Network for Papua Peace Land." *Jurnal Ilmu Sosial Dan Ilmu Politik* 21 (2): 147–159.

Pattiselanno, F., and A.Y.S. Arobaya. 2014. "The Future of Papua's Biodiversity Is Alarming." *Jakarta Post*, December 1. https://www.thejakartapost.com/news/2014/12/01/the-future-papuas-biodiversity-alarming.html.

Pouwer, J. 1999. "The Colonisation, Decolonisation and Recolonisation of West New Guinea." *Journal of Pacific History* 34 (2): 157–179.

RCA (Reality Check Approach + project team). 2015. *Education Study in Tanah Papua*. Jakarta: Effective Development Group: 65.

Republika. 2017. "Angka kematian ibu dan balita di Papua menurun" [Mortality Rates for Mother and Infant in Papua Decreased]. (in Bahasa Indonesia). *Republika News*, April 29.

Resosudarmo, B.P., J.A. Mollet, U.R. Raya, and H. Kaiwai. 2014. "Development in Papua after Special Autonomy." In *Regional Dynamics in a Decentralized Indonesia*, edited by H. Hill, 433–459. Singapore: ISEAS Publishing.

Ruhyanto, A. 2018. "Papua: How Indonesian President Jokowi Is Trying—And Failing: To Win Hearts and Minds." *The Conversation*, December 5. https://theconversation.com/papua-how-indonesian-president-jokowi-is-trying-and-failing-to-win-hearts-and-minds-107826.

Sloan, S., M.J. Campbell, M. Alamgir, J. Engert, F.Y. Ishida, N. Senn, J. Huther, and W.F. Laurance. 2019. "Hidden Challenges for Conservation and Development Along the Trans-Papuan Economic Corridor." *Environmental Science and Policy* 92: 98–106.

Smith, M. 2018. "West Papua Liberation Army Claims Responsibility for Papua Killings." Radio NZ, December 5. https://www.rnz.co.nz/international/pacific-news/377548/west-papua-liberation-army-claims-responsibility-for-papua-killings. Accessed 23 July 2019.

Syailendra, E.A. 2016. "Inside Papua: The Police Force as Counterinsurgents in Post-Reformasi Indonesia." *Indonesia* 102 (1): 57–83. https://doi.org/10.1353/ind.2016.0019.

Timmer, J. 2005. "Decentralization and Elite Politics in Papua: State, Society and Governance in Melanesia." Unpublished Paper. The Australian National University (ANU), Canberra.

Upton, S. 2019. "The Impact of Migration on the People of Papua, Indonesia. A Historical Demographic Analysis." Doctoral Dissertation, Department of History and Philosophy, University of New South Wales.

Viartasiwi, N. 2018. "The Politics of History in West Papua-Indonesia Conflict." *Asian Journal of Political Science* 1–19. https://doi.org/10.1080/02185377.2018.1445535.

White, G. 2007. "Indigenous Governance in Melanesia." State: Society and Governance in Melanesia, No. 5.

Widjojo, M.S., ed. 2010. *Papua Road Map: Negotiating the Past, Improving the Present and Securing the Future*. Jakarta: KITLV, LIPI, and Yayasan Obor.

Yunifiar, M.Q., T. Kotaki, A.M. Witaningrum, S.Q. Khairunisa, D.W. Indriati, M. Meilani, T. Yeheskiel, S. Ueda, N. Nasronudin, and M. Kameoka. 2017. "Sero and Molecular Epidemiology of HIV-1 in Papua Province, Indonesia." *Acta Med Indonesia* 49 (3): 205–212.

Chapter 17
In These Troubled Times, Could Every Classroom Become a Site of Transformation? The Story of the SENS Program

Theodore Mayer

Introduction

This paper begins with the premise that we can *both* respond directly to the unprecedented crises currently engulfing the world *and* live beautiful, enjoyable, and meaningful lives at the same time. How will we do this? I argue we will need to marshal all the resources available to us to bring a resolute kindness and mutual respect into the communities we find ourselves in or that we create at this time,[1] *and* we will have to be unflinching in identifying and fighting at every level the deeper systemic causes of our multipronged crises.[2] Kindness, respect—and by extension solidarity—are essential because "unprecedented" means no one has faced the scale or the kind of dangers we face now, and we will not easily know what to do. They are also necessary because our self-doubt, sense of powerlessness, fear of others, and isolation are among the root causes of the crisis. Clarity and courage in the fight against systemic oppression, inequality, and greed are necessary because the stakes are so high. They are also necessary because our complex worldwide systems and our cultural training create a form of inertia that leads us to assume things will go on as before, and therefore that we already know what to do. We now know enough about the climate

[1] Principled kindness and respect can and must be extended based on common humanity to all members of the global community, even to those most responsible for the crisis. They will need to be prevented from doing further harm but also invited to abandon their privileges and join the rest of the human race in working on solutions.

[2] Anthropologist Jason Hickel's thoughtful critique of the SDGs is simply summarized in this statement: "The UN's new Sustainable Development Goals aim to save the world without transforming it" (Hickel 2015).

T. Mayer (✉)
Institute for Transformative Learning of the International Network of Engaged Buddhists, Thailand
e-mail: director@inebinstitute.org

M. Indrawan et al. (eds.), *Civic Engagement in Asia*,
https://doi.org/10.1007/978-981-16-9384-7_17

crisis to say that things will not go on as before (McKibben 2019).[3] This means we will have to work together toward never-before-contemplated solutions.

One such solution is the School of English for Engaged Social Service (SENS) that I was commissioned to design in 2015 by the International Network of Engaged Buddhists (INEB). After seven years SENS is now the primary place where I explore and empirically test how to cultivate the qualities and knowledge that young adults need to lead with integrity.[4] My aim in this paper is to share in outline form what we have learned and to encourage others to turn their classrooms into hopeful communities aiming for a far-reaching transformation. This can be done whether one is designing a course and classroom from scratch, or simply introducing modifications into an existing curriculum.[5]

Transformative learning may be legitimately understood in its narrow sense as something new and different that takes place within the sphere of the classroom, aiming for changes in attitude, capacity, or worldview of the students. The small group of scholar-activists who proposed this volume, however, have allowed for a much broader sense of transformative learning as something that takes place within civil society as a whole, as a kind of civil learning. The SENS program stands roughly in the middle of this range of worthwhile approaches. It moves flexibly between the classroom and the "field" of civil society, and in this, it follows the anthropological tradition. Unlike anthropology, however, it aims to cultivate leadership that understands and can respond to the crises facing humanity at present, and in this sense, its aim is transformative at a very broad level. Put more precisely, it aims for a deep transformation of individuals in small groups as a path to a broader transformation of civil society.

Background to SENS—INEB's Intention to Foster a New Kind of Higher Learning

One tendency in the interpretation of Buddhist traditions in the modern period has supported the need for personal *and* social structural change so clearly that there is a widely used English term for it, namely socially engaged Buddhism. The late Vietnamese monk Thich Nhat Hanh and the Dalai Lama are two of the most well-known

[3] There are numerous references one could cite. For simplicity and accessibility, I suggest that readers look into this recent work by pioneering climate activist Bill McKibben for a single and very recent summation of the conclusions and concerns of climate scientists.

[4] When I say "empirically test" I do not mean double-blind comparisons. I am speaking of the day-to-day and often overlooked work of every sincere teacher. That work consists in observing and noting changes in their students in many realms and making regular assessments of what works and what doesn't to facilitate growth not only in discrete skills but also, if we are serious about transformation, more diffuse qualities such as courage, kindness, and willingness to take on responsibility.

[5] In either case, we will thus expand the community of practitioners who can share discoveries about how to realize change in the classroom and beyond.

Buddhists of the late 20th and early 21st century who fall within this relatively recent tradition. The International Network of Engaged Buddhists (INEB), co-founded in Thailand by Sulak Sivaraksa, Pracha Hutanuwatr, and others in 1989, is a non-sectarian, face-to-face network within this tradition. Like socially engaged Buddhist movements elsewhere, it was a response to major forms of suffering in the national context—life under military rule, gross inequality, the exploitation of farmers and the rural poor, rapid ecological destruction, and the growing corruption of the Thai Buddhist Sangha.

I came to Thailand in 1998 as a graduate student to study the kind of Buddhist thought and practice created by Thai socially engaged Buddhists. I focused on a new form of contemplative action inspired by Cambodian monk Maha Ghosananda and his colleagues—the Dhamma Walk.[6] In Thailand, such walks developed with a strong focus on the connection between Buddhist principles and care of the environment. The first Thai walk circled Songkhla Lake in Thailand's south over 30 days in 1996. The mixed walk contingent of from ten to more than a hundred monastics and laypeople[7] met with local communities along the way and studied through direct personal encounters the destruction of the lake's ecosystems, the welfare of lakeside communities, and the role of Buddhist monks and laypeople in bringing about change.[8] When I joined for the first time in 1997, I saw the Songkhla Lake walk as a vivid and inspiring form of experiential learning, engaging the mind, the body, and all the senses. The change brought about in the individual walkers' sense of their relationship to nature and the social world encountered in lakeside communities was palpable.[9]

Accepted as a lecturer at the Thai branch of a U.S. university in 1999, I continued to participate in such walks for many years, including the Dhamma Walks led by Phra Paisal Visalo in the mountains of Chaiyaphum, walks that continue to this day. In this way, I came to know Sulak and other leaders of the broader movement of socially engaged Buddhists in Thailand quite well.[10] Teaching at a U.S.-based university so

[6] See "Step by Step from Cambodia to ASEAN" in this volume. The "Dhamma" in "Dhamma Walk" is the Pali-derived spelling used in Theravada Buddhist contexts such as Sri Lanka, Myanmar, Thailand, Laos, and Cambodia. "Dharma Walk" would be the Sanskrit-derived equivalent used in Mahayana Buddhist contexts such as China, Japan, and Korea or Vajrayana Buddhist contexts such as Tibet, Mongolia, and Bhutan.

[7] Phra Maha Ghosananda himself spoke to walkers a few days into the very first Dhamma Walk for Songkhla Lake in 1996 (Mayer 2005, 269n17). Over the years the walk brought together local villagers, who sometimes walked in large numbers to the next village or beyond, with writers, artists, scholars, students, activists, adventurers, and well-known monks not only from the south but also from around Thailand and at times from around the world.

[8] The walk also made a point of visiting Muslim communities and addressing Buddhist-Muslim relationships when possible.

[9] This assessment is based on my interactions with and observations of walkers over many walks, as well as on the responses of walkers to questionnaires given out by organizers at the end of a given walk or when walkers had to leave (Mayer 2005).

[10] In fact, the walks brought me into contact with a rich and diverse community of writers, practitioners, educators, activists, and community leaders from around Thailand. Their skills in integrating meditation, activism, social critique, and innovative approaches to learning were often deeply rooted

close to this broader context also allowed me to take students to meet and sometimes study with the Thai leaders and thinkers I had met in my ongoing field research. It was for this reason that in 2014—facing a personal and professional crisis—I felt comfortable approaching Sulak at an event in Bangkok to ask half-jokingly if I could work for him. He surprised me by saying that he and his colleagues were hoping to start a new Buddhist-based alternative university. Perhaps I could help work on the design and implementation of programs. I began working for INEB in August of 2014, but in spite of hard work over two years, our proposal for an MA in Socially Engaged Buddhism, with field study in Thailand, India, and Taiwan, could not find a stable partner accredited within the Thai higher education system.

By the middle of 2015, however, plans for another program had emerged from our newly formed "INEB Institute."[11] Core leaders had decided that a three-month English program would help prepare our constituency of young Asian NGO workers and activists for the MA program we had envisioned. So I set to work on designing and organizing the new program. By the time our first English program had begun in February of 2016, Pracha was well on his way to designing a six-month program initially called Buddhist Leadership Training but later changed to Awakening Leadership Training or ALT. ALT and our English program (later called SENS) shared several features: importantly, both sought to realize a form of education that would address not only the "head" (the intellect), but also the "heart" (emotional connection and deepening wisdom), and "hands" (practical application and learning through doing).[12] Both programs sought to cultivate leadership. Both stressed self-knowledge through contemplative and interpersonal practices. And both took students into the field to learn from local communities and leaders.[13]

They differed in that ALT was designed as a series of modules that could be studied in their entirety or attended selectively.[14] SENS was designed to be a single stable community that could grow together for the entire period of three months under the close supervision and support of a single team of teachers and tutors. Of course, all SENS curricula were integrated with the need to develop English-language skills, while ALT assumed them. This also played a part in drawing on different constituencies: SENS drew on promising but often marginalized young leaders from Asia, while ALT attracted this group but also educated seekers with a countercultural

in Thai Buddhist culture, yet many of these individuals were open and eager to explore compatible and workable approaches from elsewhere.

[11] INEB Institute is short for the Institute for Transformative Learning of the International Network of Engaged Buddhists.

[12] "Head, heart, and hands" is a slogan evocative of Thai Buddhist movements for educational reform, but it is only the most visible sign of Thai currents in alternative education with a deep and complex history. For a relatively recent example, see Sulak (2009, chap. 4).

[13] Learning from local communities and experience in the field has also long been a hallmark of INEB's biennial conferences, which are hosted by an engaged Buddhist community in some part of Asia, and which the international participants have the chance to learn about through exposure trips and interviews outside the conference venue.

[14] This also allowed for the possibility of bringing in a wider range of primary teachers depending on the needs of each module.

bent from around the world. Importantly, SENS and ALT learned from each other's approaches, successes, and failures, and as sister programs provided for ongoing work or learning opportunities for graduates of either.

Integrity, Connection, and Play—Guiding Aims of the SENS Program

The remainder of this paper focuses on the lessons of four years of the SENS program, a program of which I have intimate knowledge.[15] The three guiding terms of our complex curriculum are integrity, connection, and play. Since the entire course came into being as an attempt to respond with integrity to the dangers of the crises facing us, we must begin there.

Integrity

Integrity is a beautiful English word with a variety of meanings. These include the ability to maintain the coherence of one's being, the quality of being incorruptible, and many variations of "being true to" something. For example, true to oneself, true to one's principles, true to one's word, and true to the needs of a given situation. The last of these is especially germane to the integrity of leaders. Suppose, for example, that hundreds of people were riding a train, and the conductors received reliable information from those entrusted with the study of track conditions that a bridge spanning a huge chasm thirty kilometers ahead had collapsed. Would the conductors not do everything in their power to alert the driver as well as the other passengers? And were they to nonchalantly go about their business and fail to alert anyone, would we not find them lacking in integrity, not to mention basic humanity, especially in light of their positions of responsibility?

I mean this of course as a metaphor for the climate crisis. The train is the earth and the collapsed bridge is the consequences of allowing global temperatures to rise more than 1.5° C above pre-industrial levels. At 1° C of extra heating above pre-industrial levels, we already have a foretaste of these consequences in the form of drought, heatwaves, crop failures, flooding, ocean acidification, sea-level rise, and accelerated species extinctions (McKibben 2019).[16] The IPCC warned in its 2018

[15] That Sulak welcomed me to work within the guidelines of the INEB Institute, even though I am a U.S. citizen raised in a Christian missionary family and inspired as much by European philosophers as by engaged Buddhists, is a testament to his openness and his lifelong commitment to provide opportunities to those younger than he, whom he feels can offer something of value if given the chance. Sulak has likewise encouraged the development of the abilities and leadership of large numbers of people in Thailand and beyond, especially young adults.

[16] Average global temperatures have by now risen roughly 1° C over pre-industrial times. Over the last 200 years, concentrations of CO_2 have gone from 275 to 400 parts per million, and they may

report that keeping within 1.5° C will require "rapid and far-reaching transitions" that include a worldwide shift to renewable energy, dramatic increases in energy efficiency, reductions in consumption, expansion, and protection of natural carbon sinks through reforestation and other approaches, and widespread changes in agriculture, industry, and infrastructure. The scientific consensus is that if we do not accomplish these changes by 2030, we will face catastrophic consequences.[17]

Climate change is a difficult issue because it is unprecedented, and because the impacts of the unconstrained burning of fossil fuels are comparatively slow. Yet for no issue except the accidental or deliberate use of nuclear weapons are the stakes so high. On these grounds, we can argue something simple here, and to students of SENS. A leader at any level of the social order, even parents, must be aware of the basics of climate change and must decide how they will participate in bringing it under control. We provide our students with the chance to study graphs and charts, read, and watch videos that detail the rapidly accumulating scientific data on climate change. Yet in asking ourselves and our students to take on a life of integrity in the sense of "responding to the needs of the situation," we face the fact that the climate crisis is not a simple or isolated one; indeed "the situation" is exceedingly complex.

Climate change persists as a difficult social problem because of huge power inequalities and vested interests, and an integrated world political economy designed to serve those interests above all. The inequalities are exacerbated by social divisions and oppression, by undemocratic decision-making (particularly in the economic sector), and by the willingness to use violence as a tool of policy.[18] Furthermore, individuals do not stand apart from these problems but are deeply implicated in them.

What we begin to see then is an array of interlinked crises, each of which maintains and prolongs the other. Our response in SENS has been to take these tightly woven crises and allow them to form the skeleton of our curriculum. We depict them as in Fig. 17.1, with the broadest and most threatening to civilization as a whole at the top to the most immediate and personal and thus most destructive of leadership and coherent action from the grassroots at the bottom.[19] Following the tradition of

---

well reach 700 ppm if there is no course correction (McKibben 2019, 21). Just three years on, in 2022, we are obliged to modify McKibben's statement to say that we have gone from 275 to "well over" 400 parts per million in roughly 200 years. McKibben adds, "Since none of us knows what a 'part per million' feels like, let me put it in other terms. The extra heat that we trap near the planet because of the carbon dioxide we've spewed is equivalent to the heat from 400,000 Hiroshima-sized bombs every day, or four each second." (2019, 21–22).

[17] A less well-known but very alarming dimension of the climate crisis is that the oceans have absorbed 93% of the extra heat produced by the burning of fossil fuels so far; furthermore, carbon entering the oceans has increased their acidity by roughly 30%. The combination of continued warming and acidification could make the ocean inhospitable for virtually any marine creature, having huge impacts on the food chain (McKibben 2019, 47–50).

[18] While one could cite many examples, perhaps closest to the climate crisis are the reports of large numbers of murders each year of environmental and indigenous activists defending forest land from destruction (Brown 2019).

[19] I hold that authentic leadership arises from within and requires clarity and confidence in one's values and views. Yet it is common to encounter to varying degrees, as I have among my students, feelings of isolation, fear of different others, doubt about their capacities, and striving to prove

- Climate Crisis or Breakdown
- Violence as a major Tool of Policy and the Background Threat of Nuclear Weapons
- Frustration of Attempts to Realize Genuine Democracy – Lack of Voice

↑ *Involves primarily a study of the external, structurally present elements of the crisis*

- Severe Social Inequality, Social Division, and Associated Oppressions
  → *Both external & experiential study*

↓ *Involves primarily a study of the experiential and internalized elements of the crisis*

- The Embattled, Disconnected Self – Feelings of Fear, Isolation, and Inadequacy in the Face of Neoliberal Standards of Respectability and Virtue

Fig. 17.1 Key areas of study within the SENS curriculum

engaged Buddhism, we study both the external and the internal manifestations of these elements of the crisis.[20]

The curriculum begins by attempting to meet and to encourage each student where they are on the first day. Thus, it begins at the bottom of this schema and moves gradually to the top. However, we never leave behind our focus on the student. At each stage, we present students with the chance to rethink their relationship to

themselves using external standards such as wealth, status, educational background, and position. Such attitudes and standards are constantly reinforced by social and mass media messages and need to be understood as primarily the internalized effects of oppressive systems. As such, those who hold them can hardly be blamed. Yet they must be guided toward critical reflection on those systems' impacts if authentic leadership is to take root. For the more we are under the sway of those messages the less likely we are to embark on leadership that requires (a) working on behalf of the marginalized or (b) questioning the very standards that rank one's self or the marginalized as failures. In this sense, such systems are highly destructive of individual potential. I see sustained leadership from the grassroots arising only when such feelings and conditionings can be reflected upon and openly questioned, making room for new attitudes toward self and one's capacity for leadership.

[20] In fact, all elements of the crisis are both external and experiential. Thus, with regard to the climate crisis, we ask students to consider talking about their feelings for nature and the grief they may experience at the loss of some part of nature they love and relate to. It is still true that certain elements of the curriculum, those at the bottom of the list, are intrinsically and primarily personal.

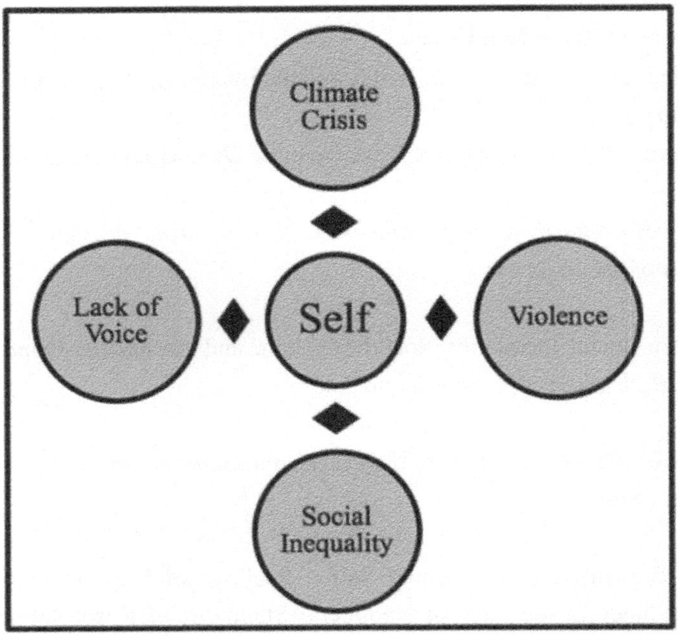

Fig. 17.2 Contemporary interlinked crises

each crisis at ever broader levels. Thus, the fulcrum of the course is the point at which the individual student meets and takes up each element of the crisis. This includes the students facing themselves and their traumas and overcoming barriers to appreciating the selfhood, joys, and traumas of the others who form the learning community (Fig. 17.2).

Connection

There is a great difficulty that prevents us from beginning the course with, say, the climate crisis. And that is that students entering our course bear the scars of oppression and disconnection in multiple ways. I would argue this is not because many are from marginalized communities. Rather, like all of us, they live within a deeply hierarchical, consumerized, mass culture that breeds disconnection in everyone.[21] Many are to varying degrees disconnected from their own minds and from knowing what they want, rather than what their parents, the mass media, or their workplace say they should want. They are often disconnected from the details of their own

[21] A similar argument is at the heart of Johann Hari's extended discussion of the roots of depression in various forms of disconnection, in *Lost Connections* (2018).

personal journey, and from the infinite value of their unique experiences and hard-earned knowledge. Like all of us, they bring with them fears about those who are different, at least to some extent, and they introduce these fears and prejudices into the classroom.[22] Some are addicted to another form of disconnection, namely the dense but largely superficial engagements of social media. Many are completely disconnected from nature and natural environments.[23] If students enter our course in this way, how can we expect them to care about the climate crisis and to believe that they could make a difference? Disconnection from their own social power, regardless of social position, is another dimension of the problem. Were we to enter immediately into the climate crisis and the urgent need to do something, many students would respond, but the response would be weak and temporary if it were built upon these many forms of disconnection. It would become another "*should*" that they would need to integrate into a life often already filled with multiple "*shoulds*."

The very first thing we do to address disconnection is deliberately build a very diverse student body while selecting only those applicants who appear ready to encounter different others with genuine curiosity and respect. As a result, when a Buddhist student enters our learning community, they will meet Muslim women wearing the hijab, and Christians from Myanmar or a tribal community. They meet other students from diverse linguistic, religious, class, and national backgrounds. All these students come together to learn English, and English becomes a zone of commonality, safety, and neutrality.[24] It is something one can always focus on, and one joins together with all the others in doing so…in spite of fear, in spite of prejudice, in spite of isolation.

Our instruction in basic forms of meditation, and the invitation to meditate briefly at the beginning of every morning class, combined with workshops and field trips that may include extended contemplative practice, opens up the opportunity for the student to be with his or her mind and body, and to feel relaxed about doing so in the company of others. By engaging in such practices, students come to pay attention to and value their own inner experience. And they see that it has merit and that it is acceptable socially to do. This is one of our approaches to disconnection from self.

Another way we show our respect for the mind and the unique experience of each student is to teach them to take turns listening to each other as they talk about life or any thoughts and feelings they have.[25] In their turn as the speaker, they are challenged to think about what they are willing to share with or show another (with agreements of

---

[22] As a result of the severe oppression experienced by students from Dalit and tribal backgrounds, successfully making it to the course may in itself be an arduous and heroic effort.

[23] This is least true for the sons and daughters of farmers, who make up a significant minority of our students.

[24] In this way the English focus of the curriculum is strangely similar to the breath or other objects of focus in the meditation practice of many traditions, to which one returns as a matter of discipline, and as a response to losing one's way.

[25] In our course we refer to this process as "co-counseling." It is, in fact, an adaptation to the needs of the situation of the theories and practices of the world-wide community of peer counselors known as Re-evaluation Counseling. The best entry point for a study of this community's history, ideas, approaches, and publications is the following website: https://www.reevaluationcounseling.org. See

confidentiality); they gradually learn that their courageous and free choice of what to address, and their way of communicating, including the expression of emotions, are all accepted and respected by their listening partner. We also teach through detailed demonstration and practice what it means to truly listen to another with respect, to refrain from giving advice, and to see whenever possible the beauty and the strength of the person speaking. This is a core component of the SENS course, without which the course would be much different, in my view.[26] This is because it introduces a very personal kind of challenge, which when met leads to the ability of the students to be both vulnerable and gentle in their relationships with each other. Teaching these listening practices, we open a safe space for the expression of emotions, yet the safety of that space is only ensured by listeners who learn to respond with kindness in a number of clearly demonstrable ways.[27] The courage of those in the speaker's role to show themselves and to express feelings, and the intention of those in the listener's role to respond with love and kindness, ultimately allow genuine emotional growth to take place for both.[28]

There is another way we address disconnection very directly, and that is to look broadly at the workings of oppression and the deliberate separation of peoples worldwide, such that one group is targeted for systematic mistreatment by the other. Racism, classism, ethnic oppression, sexism, and even language oppression are all examples of these processes. During the first half of the course and beyond, we devote a week or more to exploring each of these various forms of oppression. We do so primarily in three ways. First, we introduce the students through lectures and discussions, readings, short videos, and films to historical and current instances of a particular kind of oppression, such as racism.[29] Second, we invite students to talk about key social issues in their country, either as individuals in a mixed group or as members of a panel from that country. At times the question is left open, and at other

---

especially the article entitled "How to Begin Re-evaluation Counseling" (Re-evaluation Counseling 2021).

[26] Having included short listening sessions not only in language classes but in other academic subjects for many years, I would never want to do without this approach to learning. Inevitably students' attention in the full class will wane, especially if the material is difficult. Having the chance to break the rhythm of the class and tell someone privately what they are thinking or feeling is a powerful way to bring their attention back to life, often returning the class to lively engagement. For this reason, I recommend it heartily to other educators; this will require learning how to run such sessions, but any educator with a genuine interest can do so.

[27] That they are demonstrable allows them to also be openly critiqued and reflected upon in the learning community, with the result that genuine learning can happen in this area too.

[28] Indeed, we find that when students are free to express emotions through laughter, tears, shaking, animated talking, or even yawning, we begin to see deeper attitudinal changes, and the thinking of students appears to become clearer and more "in-touch" with themselves and their surroundings. Our findings in the SENS classroom are in line with the observations of the co-counseling communities over many decades.

[29] We show outstanding examples of dramatic films in English every Wednesday night. Due to their often disturbing and heavy content, we show documentary films now only in the afternoons when they are needed to move the curriculum forward.

times we invite a panel of women, say, or Dalits, to talk from their personal experience about what it has been like to be from that group. Finally, the issues of social mistreatment and prejudice invariably come up within the listening or co-counseling sessions.[30]

We find that the documentaries and readings on instances of oppression far afield from the experience of our students give them a chance to see that they are not alone, either in their home community or in their national context, with that kind of oppression.[31] Such films and readings give students a larger historical view, and they can begin to conceive of oppression and huge power imbalances as a feature of the larger social world. They can also begin to see each person's placement within systems of oppression as something no one should be blamed or humiliated for.[32] Furthermore, in seeing Martin Luther King, Jr. or Yazidi leader Nadia Murad respond to oppression, they understand more deeply how ordinary people can become extraordinary leaders, and they start to have a more vivid sense of the social power of those normally considered powerless. It is not a big step then to consider themselves as potentially also powerful.

Disconnection from nature is something we also address through films, documentaries, and readings, especially as we begin to study the climate crisis. Also, the course takes place in a natural and rustic setting, and our field trips take the entire group to mountain forests and the seashore. In SENS 2019 we introduced an activity at a forest monastery where, deep in the forest, students and tutors split up into pairs; one person was blindfolded and led through the forest by the other to experience trees, rocks, and various things through touch, smell, hearing, and so on.[33] This was a memorable experience for many of the students, especially those from urban backgrounds.

---

[30] These issues sometimes come up in private sessions of two or three people, but at times also in sessions with the whole class, where listening to an individual talk about their experiences of oppression is demonstrated as a way to assist students to listen thoughtfully and effectively.

[31] For example, in our most recent course we showed the documentary *On Her Shoulders*, about the struggle of Nobel Prize awardee Nadia Murad to bring the genocide of her Yazidi people, and the use of sexual violence in this genocide, to the attention of the wider world (Bombach 2018). The film is moving and personal, and it demonstrates the struggles of someone who became a leader without their intending to do so.

[32] We argue that those enacting the oppressive side of a relationship only do so based on repeated and often brutal training as children, which is ultimately a form of trauma. This opens the way to seeing them as members of a common humanity and opens the way to observing how every individual may at times take on an oppressive stance. Seeing oppressors as victims of a harsh societal training does not absolve them of responsibility for their current and future actions.

[33] We are grateful to Soeui Fah of Hong Kong, Assistant Director for the SENS 2019 course, for introducing this activity into the SENS course. Her innovation is an example of a broader process in which students, as well as work team members, introduce new ideas and solutions to challenges that then become a regular part of the course. In fact, each participant in the course has contributed much to our knowledge about what works and what engages the learning community.

Play

In the mid-1980s, I started a small Spanish language school in St. Louis, Missouri, mostly serving people concerned about the militaristic and anti-democratic nature of U.S. policy in Central and South America. Teaching Spanish initially as I had been taught, I soon felt limited by those methods and began to study alternatives in earnest.[34] I found a rich literature, with practitioners in the U.S. and Europe experimenting with new language-teaching approaches.[35] Central for me was the Natural Approach, which held that language acquisition took place almost effortlessly through the students' willing and interested engagement in communicative activities using the target language (Krashen and Terrell 1983). Teaching *about* the language through formal grammar instruction could help students become successful "monitors" of their language, especially in writing. But acquisition leading to fluency was about playful, relaxed, and interested engagement. The "relaxed" part was central, as worried or anxious students would not have sufficient attention to absorb the language through classroom interactions.

The Natural Approach, combined with Total Physical Response (Asher 1977), Accelerated Learning (Rose 1985), and theories of multiple intelligences (Gardner 1985) led me to begin creating a classroom that engaged the students through art, color, movement, games, stories, and more. Over time my language classroom had become, in a word, playful. I found this had a dramatic effect on the students' enjoyment and depth of learning. Yet after several decades of using such approaches, especially with students from developing countries, I concluded that one source of student anxiety needed to be addressed more explicitly. Students tended to enter a classroom for the first time with this kind of (unspoken) question: "What must I do to demonstrate in this particular classroom that I am an intelligent and competent person?"[36] Two of the common answers are: (1) Don't speak up unless you are quite sure you know the correct response, and (2) Don't make any mistakes if you can avoid it.

In my view, the underlying issue is the ranking of intelligence and skill that is so widespread that many students feel it is, in fact, the central function of educational institutions. In this, they may not be far off the mark, as they know that the world economy—and that of their society—offers desirable jobs and lifestyles for an elite few while leaving the rest to struggle for a decent wage, livable working conditions, and sometimes for any form of employment at all. That the students think in this way is a perfect reflection not of social life but of social life under the neoliberal order. For me, the widespread assumption in academia and elsewhere that some are dull

---

[34] My language teachers were at the same time the source of wonderful ideas and practices, some of which I continue to use to this day.

[35] It was interesting to me when I attended conferences that while many of the leading writers and practitioners held university positions, many if not the majority of those trying the new approaches were doing so in small institutes and enterprises outside of formal academia.

[36] For many students the question would have to be framed in stronger terms: What must I know and do here to be certain I will not appear foolish or in any way bring humiliation upon myself?

while others are brilliant is one of the most damaging assumptions a teacher could make. I hold instead that every one of us is highly intelligent, but that each one of us faces the life-long challenge of overcoming all obstacles to the *cultivation* of that intelligence.[37]

We begin our SENS programs on the first day by explicitly stating our view that each individual in the classroom is blessed with abundant intelligence, but that we all face the challenge of learning to use it. We also affirm explicitly that mistakes are an important kind of feedback for our learning, and at the beginning of the course, we ask everyone to applaud when someone has made a mistake, to help drive home the idea.[38] To further fight the deeply ingrained expectation of ranking and criticism, we train the students in the first few weeks in verbally appreciating each other in English, and we rely on this training throughout the course: "I like the way you…" and many other variations.[39] We make every sequence inside and outside the classroom as engaging as we can, and it makes learning fun not only for the students but for the teachers and tutors as well.[40]

Leadership

When students can enjoy their mind at work, they are well on the way to learning that is rapid and deep. It is a significant but not overwhelming leap from learning to enjoy the play of one's mind, to having the courage to state an unusual opinion or to question an idea expressed by another student or the teacher. The play we cultivate in this way is closely bound up with our cultivation of the students' sense of their potential as leaders. This begins by asking them to think about what they would want to accomplish in their lives and to start writing these goals down. While writing life goals is a serious task, the key here is not obligation but what the students would find joyful and meaningful. And that is a question of play, if we take the word a bit seriously.

---

[37] There are many reasons to question the entire intelligence ranking apparatus. However, that argument is beyond the scope of this paper. I refer the reader to another paper in which I address this issue in somewhat more detail. See Mayer (2018, 1–39).

[38] In some years, this practice catches on so well that it continues throughout the course, including for mistakes by the teachers and the work team, which the students are especially delighted to acknowledge in this way.

[39] Our students become very adept at offering verbal appreciations of others. After every workshop or lecture by a guest speaker, we propose having a round of appreciation from the students. Some of the invited leaders say it is not necessary, but once they have heard the insightful appreciations from students they are often delighted and moved.

[40] In fact, this is one of the main types of ongoing work that we do to make the course function well. From the point of view of the natural approach, this is essential if one is to foster effective language learning. Furthermore, when the joy of learning is matched by the joy of teaching, the entire enterprise becomes socially sustainable in a way that is not always true for the teaching profession.

To further cultivate their sense of leadership, we do the following: (1) ask them what they think on several issues; (2) ask them what they would want for their lives and their community if it were up to them; (3) take them to meet outstanding Thai leaders from diverse backgrounds[41]; (4) ask them to research and give a presentation on an outstanding leader[42]; (5) make vivid and inescapable the urgent need for leadership in responding to our contemporary crises[43]; and (6) expose them to numerous examples of hopeful and constructive action in the world.

In the recent period, our Myanmar students have helped us to learn how we can support their leadership under entirely new conditions. The military coup of February 1, 2021, has been devastating for Myanmar society, and for every one of its ethnic groups. The coup regime has forced civil society to go completely underground, as activists are targeted, detained, and frequently killed. In that context, we have begun to offer to small groups of Myanmar civil society activists the same SENS program described in this chapter. There are two important differences from our previous programs: first, we must offer them in a fully online format, and second, in doing so we are engaging with young adults who have already embraced a life of leadership that demands remarkable courage. Their commitment to lead has involved them in a wide variety of activities, including: sending supplies to villagers who live in the forest to escape the bombing of their homes, sharing knowledge of sustainable farming and health practices, and supporting the democracy movement. All of this they do at great personal risk; at least once a week a few of the students are absent because conditions of conflict and surveillance have forced them to move to yet another safe house or handle another emergency. In this context, we do our best to offer what one student called "an oasis" of mutual support, exposure to the outside world, and the joy of a regular study environment to young people who are, in many respects, already extraordinary leaders.

Conclusion

Cultivating kindness, mutual understanding and support, the joy of learning, and the aspiration to lead are exciting and challenging kinds of transformative work in the classroom. However, they all come down to specific attitudes and actions that the work team can take, and to the careful orchestration of these within the classroom and within the course as a whole. Our own experiments in doing these have been a

---

[41] As the students' interview and come to understand the aspirations of Thai leaders and witness the communities or movements they have founded, they also come to understand environmental issues and forms of social oppression and exclusion through the lens of that leader's involvements in the Thai context.

[42] While most students choose world-renowned leaders, some choose outstanding individuals little-known outside their national context, thus enlarging the range of examples of leadership that students can study and emulate.

[43] This also includes facing the fact of relative inaction and sometimes downright obstruction from those who claim to be political leaders.

mixture of successes and failures. But how could it not be so if learning invariably involved making mistakes?[44]

Based on our experience, the attitudes and actions necessary to lead this kind of course are accessible to anyone with sufficient commitment and willingness to learn from the available literature, from their own and others' successful experiments, and their own and others' mistakes.[45] We feel we have demonstrated that the transformation of the language classroom into a vehicle for wider social transformation is certainly possible. Our approach of cultivating awareness, sensitivity, and leadership in small groups will take time. This is not cause for despair, however. Within the intimate scale of the learning or work community, the growth in confidence, clarity, and determination to act of individuals in each SENS cohort is profoundly inspiring.[46] At the scale of larger social entities, we now know that the success of a social movement (to end the climate crisis, for example) does not require the engagement of a majority of the population, but rather around 3.5% (Chenoweth 2017). On these grounds we invite readers, especially those who are educators and policymakers, to join us in transforming your institutions and classrooms into vehicles of genuine change for individuals and for the societies that set important parameters for our lives.

References

Asher, James. 1977. *Learning Another Language Through Actions*. Los Gatos, CA: Sky Oaks.

Bombach, Alexandria, dir. 2018. *On Her Shoulders*. DVD. https://rmwfilm.org/films/on-her-shoulders/.

Brown, Alleen. 2019. "More than 160 Environmental Defenders Were Killed in 2018, and Many Others Labeled Terrorists and Criminals." *The Intercept*, July 30, 2019. https://theintercept.com/2019/07/30/criminalization-environmental-activists-global-witness-report/.

Chenoweth, Erica. 2017. "It May Only Take 3.5% of the Population to Topple a Dictator–with Civil Resistance." *The Guardian*, February 1, 2017. https://www.theguardian.com/commentisfree/2017/feb/01/worried-american-democracy-study-activist-techniques.

Gardner, Howard. 1985. *Frames of Mind: The Theory of Multiple Intelligences*. New York: Basic Books.

Hari, Johann. 2018. *Lost Connections: Uncovering the Real Causes of Depression–and the Unexpected Solutions*. New York and London: Bloomsbury Publishing Plc.

Hickel, Jason. 2015. "The Problem with Saving the World." *Jacobin*, August 8, 2015. https://www.jacobinmag.com/2015/08/global-poverty-climate-change-sdgs/.

IPCC. 2018. *Summary for Policy Makers on IPCC Special Report on Global Warming of 1.5°C*. Geneva: World Meteorological Organization. https://www.ipcc.ch/sr15/chapter/spm/.

Krashen, Stephen D., and Tracy D. Terrell. 1983. *The Natural Approach*. Hayward, CA: Alemany Press.

[44] Including all, or even our major, mistakes, would have made for a paper far beyond the word limit provided by the editor.

[45] With regard to sharing what we have learned in SENS, we are contemplating taking the further steps of (a) offering intensive training for teachers in Asia or beyond who would like to use these approaches; and (b) creating a step-by-step manual for those who would like to turn their classrooms into sites of transformation.

[46] See the pieces by Sabin Rongpipi and Mahesh Admankar in this volume.

Mayer, Theodore. 2005. "Paths to a Possible South: The Dhamma Walk for Songkhla Lake." In *Dynamic Diversity in Southern Thailand*, edited by W. Sugunnasil, 241–273. Chiang Mai: Silkworm.

Mayer, Theodore. 2018. "Thoughts on Designing Transformative Education for Buddhist Communities in ASEAN." In *Contemporary Buddhist World: Motivating & Moving Asian Countries*, edited by A. Lochan, 1–39. Bangkok: Thai Tibet Center.

McKibben, Bill. 2019. *Falter: Has the Human Game Begun to Play Itself Out?* New York: Henry Holt and Co.

Re-evaluation Counseling. 2021. "How to Begin Re-evaluation Counseling." https://www.reevaluationcounseling.org/how-to-begin-rc.

Rose, Colin. 1985. *Accelerated Learning*. Aylsebury, Buckinghamshire: Accelerated Learning Systems, Ltd.

Sulak Sivaraksa. 2009. "Re-envisioning Education." In *The Wisdom of Sustainability: Buddhist Economics for the 21st Century*, edited by A. Kotler and N. Bennet, 41–55. Kihei, HI: Koa Books.

Chapter 18
SENS and Its Impacts on Me: A Reflection from Karbi Anglong

Sabin Rongpipi

I am Sabin Rongpipi. I am from a small town called Diphu in the Karbi Anglong district of the state of Assam in Northeast India. I belong to the Karbi community, one of the major indigenous tribes of our region.

In 2018, I was selected to participate in a three-month course in Thailand called the School of English for Engaged Social Service (SENS) at Wongsanit Ashram. It was my first opportunity to travel outside my country and to learn along with other participants from different countries. It was a life-changing experience for me because I was unaware of the outside world. For three months, I witnessed myself coming out of my cocoon by learning more about the world and more about myself. I started to discover new things about myself and my inner world, and this made me stronger and more confident. I started to connect with my feelings and my desire for an empowering connection with others. These connections were fostered through co-counseling, a core practice during SENS. In daily co-counseling sessions, we practiced deep, nonjudgmental listening in pairs or small groups. It was during these sessions that I started to realize my insecurities, my childhood traumas, my judgmental mindset, and all the odd things that I had never previously recognized. Through field trips across Thailand, meeting and interacting with grassroots leaders and experiencing powerful social projects, I learned so much, expanded my imagination, and grew as a leader.

On graduation day, I presented my long-term goals to the 2018 cohort and our supporters. My ultimate goal was to start a course similar to SENS in my hometown, which would help students improve their English while emphasizing our tribe's unique culture, language, and history. We would visit places and projects to connect with others and nature, learn self-discovery, and set long-term goals as a budding

---

Rongpipi describes her experience of the SENS 2018 course that she joined as a student (see Mayer [Chap. 17]).

S. Rongpipi (✉)
Diphu, India

M. Indrawan et al. (eds.), *Civic Engagement in Asia*,
https://doi.org/10.1007/978-981-16-9384-7_18

community of engaged young people. Within a year of having set this critical goal, I am very proud to have realized it through the Heritage Learning Program (HLP). HLP is a three-month learning journey that weaves together English fluency, Karbi Studies, and personal growth. The curriculum was modeled after SENS. Our faculty consists of an intergenerational network of Karbi elders, scholars, and youth activists who are experts in different fields, from Karbi rituals and grammar to climate change.[1]

HLP started because of our collective concern for the challenges of our rapidly changing society. Each week, students are immersed in new critical topics taught by both local and visiting experts. Through field trips and excursions, students have a chance to apply key learnings and expand their comfort zones, from leading a workshop with school children to trekking in remote villages or traveling abroad. HLP revolves around building a safe, inclusive learning community that fosters trust, openness, and deep connections among participants and it is in this space that transformation occurs and empowerment takes root.

We successfully concluded the second cohort of HLP this year. There are 30 alumni in total and many of them have continued participating in HLP activities, as well as pursuing their individual goals. In the months since completing SENS, I have also had many opportunities to continue learning and growing. I was invited to the Tribal Leadership Programme in India (2018),[2] the Mindful Markets Social Enterprise Course in Thailand (2018),[3] and the Awakening Leadership Training in Thailand (2018).[4] I became certified in Mindful Facilitation for Empowerment through Gaia Education. And I had the chance to apply my new skills as co-facilitator for an international program in Vietnam called Asian Youth Empowerment: Holistic Leadership Facilitation.[5]

It has not been an easy journey for me. My strength and power come from the belief that change begins with the self, the individual. I believe that young people like me have responsibilities, and I cannot go on blaming older generations for all the ills of the world. I must try to make a difference in my small way. I believe that I can make a difference. I can address critical issues by continuing the Heritage Learning Program, to empower youth by creating a safe space and supporting them to realize their worth and capacity as change agents.

---

[1] I am especially grateful to my main mentor, Mr. Dharamsing Teron, a cultural activist, independent researcher, and former politician who focuses on preserving Karbi folklore, songs, and oral history through his writing. He is best known for a *Karbi Studies* series, which includes Volumes 1–4 (Teron 2011, 2012; Teron and Tisso 2012; Teron et al. 2014). He is currently working on Karbi oral history and customary law.

[2] For a description of the current (2019) Tribal Leadership Programme see: https://www.tatasteel.com/initiatives/samvaad/pdf/tlp/Tribal%20Leadership%20Programme%202019%20-%20Programme%20Schedule%20_21.6.19.pdf.

[3] See https://towardsorganicasiath.wordpress.com/2018/09/23/mindful-markets-social-enterprise-course-2018/.

[4] See https://www.awakeningleadership.net/2017-awakening-leadership-training-2.

[5] See https://www.slideshare.net/pthuyduong/asian-youth-empowerment-2019-holistic-leadership-facilitation-150134828.

My transformational experiences in the School of English for Engaged Social Service, Awakening Leadership Training, and Tribal Leadership Programme have given me the insights to recognize my role in society. I am confident in my work, knowing that a single candle can light up the darkness, and I am determined to create change by supporting other youth to realize the same.[6]

References

Asian Youth Empowerment. 2019. Holistic Leadership Facilitation. https://www.slideshare.net/pthuyduong/asian-youth-empowerment-2019-holistic-leadership-facilitation-150134828.

Awakening Leadership Training Programme. 2018. https://www.awakeningleadership.net/2017-awakening-leadership-training-2.

Mindful Markets Social Enterprise Course. https://towardsorganicasiath.wordpress.com/2018/09/23/mindful-markets-social-enterprise-course-2018/.

Teron, Dharamsing. 2011. *Reclaiming the Ancestors' Voices*, Vol. 2 in *Karbi Studies*, 2nd ed. Guwahati: Assam Book Hive.

Teron, Dharamsing. 2012. *Memories, Myths, Metaphors*, Vol.1 in *Karbi Studies*, 2nd ed. Guwahati: Assam Book Hive.

Teron, Dharamsing, and Sikari Tisso, eds. 2012. *Folktales from the Fringe: Popular Karbi Folktales in English with Karbi Text*, Vol. 3 in *Karbi Studies*, 2nd ed. Guwahati: Assam Book Hive.

Teron, Dharamsing, Kabeen Teronpi, Kache Teronpi, and Valentina Teronpi. 2014. *In Search of the Drongo and the Other Stories*, Vol. 4 in *Karbi Studies*. Guwahati: Assam Book Hive.

Tribal Leadership Programme. https://www.tatasteel.com/initiatives/samvaad/pdf/tlp/Tribal%20Leadership%20Programme%202019%20-%20Programme%20Schedule%20_21.6.19.pdf.

[6] I would like to extend my sincerest gratitude to Ted Mayer, Director of SENS, and Lucy Burriss, Assistant Director of SENS 2018, for giving me the opportunity to be a part of the 3rd cohort of SENS in 2018. I am also very thankful for the assistance I received from Ted, Nilanjana Premaratna, and Melissa Storms in editing and clarifying my writing for this essay.

Chapter 19
SENSing the Truth Amidst a Social and Environmental Crisis: Learnings and Contributions Towards Sustainable Development in South India

Mahesh Admankar

Introduction

In 2017, I participated as a tutor in a dynamic 12-week program called the School of English for Engaged Social Service (SENS) at the Wongsanit Ashram in Thailand.[1] My responsibilities included a daily one-hour teaching assignment, regular planning and review sessions with the work team, written reports on my progress, and a final review of the strengths and weaknesses of the program. My main objective in joining the SENS program was to learn about Buddhist methods of personal and social transformation. I sought inspiration and techniques to fight the effects of massive discrimination, social inequality, and environmental degradation I had witnessed and experienced in India.

When I joined the SENS program, I suffered from feelings of humiliation, regret, loneliness, guilt, resentment, and fear—as well as self-criticism and lack of confidence—arising from my experiences within an oppressive social system and exacerbated by a lack of educational and economic opportunities.

[1] See Mayer, Report on the English for Engaged Social Service Program for 2017 at: https://inebinstitute.org/reportsens2017/.

Here, Mahesh describes his experience of the SENS 2017 program, which he joined as a tutor. See Mayer (Chapter 17).

M. Admankar (✉)
Department of Public Policy and Public Affairs, John W. McCormack Graduate School of Policy and Global Studies, University of Massachusetts, Boston, USA

M. Indrawan et al. (eds.), *Civic Engagement in Asia*,
https://doi.org/10.1007/978-981-16-9384-7_19

Buddhist Influences

I was born into a migrant family with roots in the Dalit social community, a community historically discriminated against and considered "untouchable." Thus, I was well aware of my place in Indian society. As a child and growing adult, like many poor people in India, I was subjected to systematic discrimination, poverty, and lack of access to an education that could empower me and my family. Through hard work and with the support of people who believed in me, I was able to earn a graduate degree from one of India's top universities. Though I had accomplished what many marginalized do not, the childhood wounds of humiliation and fear persisted. Despite finding places where I could share my experiences of pain, those spaces were not sufficient to sustain my confidence. I had been programmed to believe that I could not be an intellectual and that I deserved neither education nor opportunities for a better life.

All of these factors resulted in the development of a conflicted personality. After unsuccessful personal and social development endeavors and a deep financial crisis, I was searching for a place to find and recreate myself. An opportunity presented itself when I participated in an international conference on social engagement and liberation hosted by the International Network of Engaged Buddhists (INEB) and Nagaloka in Nagpur, India in 2016, where I met Buddhist teachers, philosophers, and practitioners from across the world.[2]

I delved further into Buddhist teachings and found inspiration in the works of Babasaheb Dr. BR. Ambedkar (1891–1956), who influenced the current social enlightenment and sustainable development work being undertaken by institutions such as Nagaloka, Manuski Trust, and Swaeroes International in India.[3] I came to know about the SENS program through its director, Mr. Theodore Mayer, whom I met in Nagpur, and soon afterward had the opportunity to participate as a full-time tutor for SENS 2017.

Doorway to a New Way of Learning

The SENS program's unique teaching methods help participants arrive at a deeper understanding of life using the English language, which facilitated sharing, listening, self-analysis, and reflection on society. SENS allowed me to meet and learn from various leaders whose stories I found to be inspirational and courageous. Workshops on co-counseling, suffering, spiritual ecology, and power analysis were enormously motivating, and interviews with activists, such as Sulak Sivaraksa, Venerable Dhammananda Bhikkhuni, Phra Paisal Visalo, and Lodi Gyari Rinpoche on their work

[2] Nagaloka is a Buddhist training and conference center in Nagpur, and is the home of the Nagarjuna Training Institute. See https://swaerocircle.org

[3] For Manuski Trust, see http://manuski.in/ and for Swaroes International, see https://swaerocircle.org

for the betterment of society offered me profound new insights. Their dedication in the face of threats showed me what it means to be truly courageous. SENS was transformative as it gave me space to openly express and share my experiences of violence, denial, and deeply rooted personal wounds through the powerful method of co-counseling. I joined the SENS family of people who practiced the values of trust, love, respect, deep listening, and concern for a fair social and physical world.

Discussions surrounding climate change and the proximate danger of human extinction also had a profound effect on me, as SENS introduced me to people who address the crisis through a nonviolent path. I experienced an awakening at SENS working under the guidance of Ted Mayer, whose teaching, philosophy, work, support, and commitment ignited a fire in me to become a more confident leader.

Shortly after completing SENS, and through the support of partners and friends of INEB, I had the opportunity to participate in a summer program on Indigenous Peoples' Rights (ISSP) at the Center for the Study of Ethnicity and Race at Columbia University in New York City. At ISSP, I learned from and exchanged ideas with activists and scholars from across the world. I toured the headquarters of the United Nations and also traveled to the US-Canadian border to learn about self-governance and the cooperative community living of the people of Saint Regis Mohawk Reservation. The enriching and supportive learning environment at Columbia University echoed my experience at SENS.

Goals, Mission, and Activities after Graduating from SENS

With the generous support and the knowledge I received through SENS, I decided to teach English, leadership skills, and social analysis, especially with regard to the struggle for social equality and the effort to respond to climate change. I based my methodology on the transformative techniques I learned in the SENS program and focused on promoting healing as well as educating my students from marginalized communities about climate change and social inequality.

Concretely, I engaged in the following activities after the program:

1. **Utilizing Transformative Education Methods in India:** Upon returning to India, I joined a public university, and later an institution working with minority youth, as a teacher. For two years, I worked with students from predominantly poor backgrounds who had suffered discrimination on many levels. I applied the methods of deep listening, free sharing of thoughts, building a support team, teaching English, and connecting with positive stories across the world. The result was an increase in students' confidence and progress in their academics. Also, I created a social platform called Awakening Students Peace Center (ASPC), which focused on supporting marginalized youth. This involves teaching English as well as supporting students in their development of interpersonal and leadership skills. Specifically, I am working with an organization called Telangana Students Union (TSU), where we also campaign against discrimination based on caste and gender.

2. **Co-counseling for Inner and Social Healing:** I created a small group consisting of students, scholars, and social workers with whom I conduct individual sessions every fortnight both at the university and also through phone calls. We share experiences, wounds, and wisdom for deeper personal and social healing.
3. **Work on Climate Change:** Drawing from my experience at SENS, I integrated modules on climate change into the classes I taught on natural resource economics at the university. My students' immediate response was a promise to use utilities efficiently, plant trees, and educate their communities. In January 2019, I began an initiative called NEST (Network for Environmental and Social Sustainability), which consists of a small cohort of concerned scholars, activists, bankers, teachers, and farmers who advocate for a sustainable world by educating each other. I plan to expand NEST further by creating an action platform for environmental sustainability through monthly workshops and on-the-ground action. Due in part to my interest in and work on climate change, I received scholarships to learn about sustainability at the Bangkok Forum held at Chulalongkorn University in 2018 and to participate in 2019 in a program at the University of Bergen Summer Research School on Climate Change in Norway.

Conclusion

One word—marginalized—has defined who I am for others. I have fought against this label and sought avenues of knowledge, mentorship, and solidarity that enabled me to reach deep within myself to know who I am, to help others to discover their true identities, and to fight the threat of climate change. My journey began at birth when I took my first breath; however, a flame was kindled within me at SENS. I know my story will be one of a lifelong battle against social exclusion. SENS is more than an academic program—it has provided the wings to carry me from a place of inner and social blockades to become a free and caring being. It not only led me to places all over the globe but also enlightened and empowered me to take initiatives in India for a just and sustainable world.[4]

[4] I am deeply grateful to Ted Mayer and Nilanjana Premaratna for their support and guidance throughout SENS 2017, and to Sulak Sivaraksa and the late Lodi Gyari Rinpoche for helping to make SENS a springboard to broader engagements that further challenged me and allowed me to grow. Without the work and support of Dr. RS Praveen Kumar Swaero, Dr. K. Lakshminarayana, Dr. Nagaraju Gundemeda, Dr. Elsa Stamatopoulou, Dr. Vamsi Vakulabharanam, and Dr. Vijay Gudavarthy, I would not have had the courage to write about my social experiences or to keep moving beyond my boundaries. I owe them a special thanks here as well. Finally, I thank Ted, Nila, Jennifer Millett, Bhaskar, Purna, Sampath, and Melissa Storms for their kind help in providing feedback and suggestions for this essay.

References Cited

Bangkok Forum: Future Sustainable Asia. https://bkkforum.chula.ac.th/bangkok-forum
Manuski Trust. http://manuski.in/
Mayer, Theodore. 2017. "Report on the English for Engaged Social Service Program for 2017." https://inebinstitute.org/reportsens2017/
Nagarjuna Training Institute at Nagaloka. http://www.nagaloka.org/nti/
Swaeroes International. https://swaerocircle.org

Chapter 20
Sustainability and Communities of Faith: Islam and Environmentalism in Indonesia

Fachruddin Majeri Mangunjaya and Ibrahim Ozdemir

Highlights

- Leaders and grassroots activists in the Muslim world have begun gearing up to meet the world's development challenges by forming strategic fora for sustainability.
- Indonesia provides examples of how civic engagement from communities of faith can help to build sustainability.
- Indonesian examples include the mobilization of instruments of Islamic finance and the reinvention of boarding schools and mosques to support environmental objectives.

Introduction

Environmental problems have come to threaten all of humanity and the very existence of life on Earth. As a consequence, in recent decades, a growing sustainability movement has developed in many corners of the world. Environmental awareness can be motivated by various drivers such as knowledge, culture, worldviews, and teachings or norms, which underlie the attitudes and perceptions that become the bases for action (Palmer and Vinlay 2003; Nurdeng 2009; Mangunjaya and McKay 2012;

F. M. Mangunjaya (✉)
Institute of Islamic Understanding Malaysia (IKIM), Kuala Lumpur, Malaysia
e-mail: fmangunjaya@civitas.unas.ac.id

Center for Islamic Studies, Universitas Nasional, Jakarta, Indonesia

I. Ozdemir
Uskudar University, Istanbul, Turkey
e-mail: ibrahim.ozdemir@abo.fi

Abo Akademi University, Turku, Finland

M. Indrawan et al. (eds.), *Civic Engagement in Asia*,
https://doi.org/10.1007/978-981-16-9384-7_20

Mohamed 2014; Mangunjaya and Praharawati 2019). Religions can also provide an important lens for understanding human worldviews, attitudes, and behavior regarding major issues such as social and environmental change and can mobilize their adherents to pursue sustainable lifestyles and action, including biodiversity conservation (Bhagwat et al. 2011).

The influence of religion on human behavior cannot be separated from religious leaders as role models; this represents an important factor in promoting sustainability. A 2007 survey from Indonesia about trust in the Muslim community by the Center for Islamic Research and Community—Universitas Islam Negeri Syarif Hidayatullah concluded that *ulema* (religious leaders) are more trusted than even the highly esteemed President of the Republic of Indonesia himself (NU 2017). Therefore, the community's religious leaders must exemplify the Quranic values and be involved in setting an example and spearheading environmental movements. Moreover, they can provide a broad understanding of the relationship between religion, climate change, and sustainability.

The global character of environmentalism has allowed Muslim environmentalists to learn from and contribute to international environmental movements. This chapter reviews initiatives, environmental actions, and movements including Islamic finance, green boarding schools (*eco-pesantren*), and green mosques as new instruments with particular reference to Indonesia.

Calls for Sustainability from the World's Faith Communities

International environmental movements have increasingly come to recognize the importance of reaching out to communities of faith, given the role that religion plays in shaping attitudes and behaviors across many corners of the globe. Indeed, many environmental organizations have found that religious communities can play a unique and fundamental role in preserving the environment. The Alliance of Religions and Conservation (ARC 2009), for example, proposed seven key elements or forces by which religion can play a role and make it possible to make a meaningful contribution if they are involved in raising environmental awareness: (1) ownership of assets, (2) educational institutions, (3) wisdom, (4) lifestyle, (5) media ownership, (6) celebrations, (7) potential partnerships.

An important global partnership began in Bristol, UK on September 8, 2015, when 24 faith groups launched the *Bristol Commitments—Faith Plans for A Sustainable Future* for meeting the Sustainable Development Goals (ARC and UNDP 2015). In the wake of the Bristol Commitments, interfaith leaders of Indonesia, under the leadership of Prof. Din Syamsuddin of the Indonesian Council of Ulama (MUI), created an interfaith forum for the environment which was named the Earth Alert *Forum Siaga Bumi*. Prominent figures from Indonesia's executive and legislative governmental branches (including the Chair of Indonesia's General Assembly) attended the launch at the House of Parliament on 21 September 2015. Leaders from Catholic, Protestant, Hindu, Buddhist, Confucian, and Islamic faiths (including its two largest branches in Indonesia, Nahdlatul Ulama, and Muhammadiyah) participated. The platform of cooperation aims to carry out religious support within the framework of accelerating sustainable development (Bona 2015).

Environmental Commitments in the Muslim World

The Organization of Islamic Cooperation (OIC) convened the *First Global Forum on Environment from an Islamic Perspective*, in Jeddah, Saudi Arabia, October 20–23, 2000, underlining the global sustainability commitments of the Muslim world. It was a wake-up call for officials and experts from different parts of the Muslim world to discuss ways to find an Islamic space for cooperation and harmonizing positions on the protection of the environment and achievement of sustainable development (ISESCO 2019).

The momentum gathered speed in the lead-up to COP-21 of the UNFCCC[1] (Paris 2015). On August 18, 2015, a group of Muslim environmentalists meeting in Istanbul proposed the *Islamic Declaration on Climate Change*. More than 100 scholars and leaders from the Muslim world contributed to the declaration, which was officially submitted to the UNFCCC secretariat before the COP-21 in December 2015. The *Declaration* called on all countries to eliminate greenhouse gas emissions as quickly as possible, to stabilize GHG concentrations in the atmosphere, and to commit 100% to switch to renewable and zero-emission energy technologies as soon as possible on a global scale (UNFCCC 2015; Mead 2015).

Shortly thereafter, the 6th meeting of ministers of OIC member states in charge of the environment and sustainable development took place in Rabat, Morocco on October 8–9, 2015. This conference resulted in the issuance of the *Islamic Declaration on Environment Protection and Sustainable Development*, which called on all 57 member countries of the OIC to pursue green economies, work toward food security and water sustainability, eradicate poverty, and embrace energy efficiency and renewables (BGT 2015). Like the *Islamic Declaration on Climate Change*, the declaration by OIC environment ministers was also taken to the platform of the Paris COP-21, which convened in December of 2015.

Indonesian Experiences

Indonesia's Muslim community has a unique experience of grassroots action for sustainability, which includes the practices of Islamic finance, green boarding schools, and green mosques.

[1] 21st meeting of the Conference of Parties (COP 21), a United Nations Framework Convention on Climate Change.

Islamic Finance

Islamic finance is as old as Islam itself and constitutes an inseparable part of an entire economic system in the world (Çizakça 2000), providing a system that encourages and can support sustainable development efforts (Hassan 2005). Islamic financial management has a certain aim, namely, to fulfill the *maqasid shari'a* (aims of Islamic law), which specifically seeks to preserve the five essentials of human well-being: religion, life, intellect, offspring, and property. Islamic financial support is based on the real sector, prohibits usury, and promotes concrete actions with collaborative efforts, intending to promote human well-being. This is to be achieved by organizing the resources of the earth on the basis of cooperation and participation (Khan 1994).

As outlined in the *Qur'an* (55: 7), Islamic teaching recognizes that natural resources are limited and must be managed within a balanced framework. Natural resource management to alleviate poverty therefore falls within the framework of Islamic finance and economic systems. Islamic finance mechanisms including *zakat* (a key instrument for wealth distribution), *infaq* (expenditure to meet social obligations), or *sadaqah* (alms for the poor as part of one's belief in God) (Lewis 2001), can be mobilized toward environmental goals related to poverty, and can also help finance to support the purpose of sustainable development.

A central aim of the Qur'an is to establish a viable social order on Earth that will be just and ethically based (Rahman 1994). Therefore, the outstanding instruments such as *zakat* and *waqf* may provide a model to alleviate poverty and mitigate inequalities in society (Raimi, Patel, and Adelopo 2014). *Zakat* is the transfer of a certain portion of a person's wealth to those who are in need, with the additional emphasis that the wealth (and assets) should come from equally lawful sources (Mohsin 2013, Arif 2017). *Zakat* is so important that it constitutes the third pillar of Islam. In practice, *zakat* refers to the *obligatory* payment of a determinate portion of a Muslim person's lawful property, for the benefit of the poor and other groups of people enumerated in the Holy Qur'an (Adachi 2018). Although *zakat* over the ensuing centuries has been subject to changing societal dynamics (Nagaoka 2014), it has been and will continue to be instrumental in establishing socioeconomic justice in the society (Arif 2017.)

The Muslim religious endowment known as *waqf* serves under Islamic law as a nonprofit social-banking system (Mohammad 2011). There are thousands of *waqf* institutions around the world that have provided much support for:

- Education: madrasahs, Qur'anic education, and universities;
- Health sectors: clinics, hospitals;
- Places of worship: *masjid*, *musallas*, orphanages, retirement homes and charities, etc.

Islamic financing schemes such as *zakat*, *infaq*, and *sadaqah* have continued to thrive in Indonesia and have received increasing legal recognition by the secular government. Indonesia passed Zakat Management Law No. 38 of 1999, which was later amended by Zakat Management Act No. 23 of 2011. The semi-governmental Zakat Management Organization, BAZNAZ, was established on January 17, 2001.

Recent dynamics see a progressive movement, namely "reconceptualization of *zakat* from philanthropy to obligation," thus further pushing *zakat* as a good financial resource for supporting social welfare (Çizakça 2000; Adachi 2018).

Indonesia has quite a few examples of Islamic finance being put into practice to support environmental objectives. In a progressive bid to support safe water and sanitation, the Indonesian Council of Ulama (Majelis Ulama Indonesia–MUI) in 2015 released a *fatwa*[2] enabling the use of *zakat*, *infaq*, and *shadaqh* for establishing clean water and sanitary facilities. In Indonesia's special province of Aceh, which is governed by Shariah Law, youth are moving to establish "*hutan wakaf*" or a religious endowment for forests from the *waqf* funding (Hutan Tersisa n.d.).[3] The latter illustrates, again, how *waqfs* should be developed and translated more broadly as long as it concerns the principles of *maslahah* (public benefit) and follows the lines of *shari'a* provisions.

Islamic finance also recently aligned itself with Indonesia's stated goal to develop a green economy. Indonesia saw in 2018 the formal establishment of sovereign green bonds—*green sukuk*—in leveraging private finance for sustainable development (SDG Philanthropy Platform 2019). Close to $1.25 billion USD were raised to cover five years. Such a level of issuance is close to oversubscription, indicating the existence of a good market for sustainable and responsible investments. A broad range of investors, including conventional and green investors as well as Islamic finance, are now involved.

Eco-Pesantren

The involvement of Islamic boarding schools (*pesantren*) in the environmental sphere in Indonesia began to be visible as early as the 1980s. This was when the Annuqayyah Islamic Boarding School Guluk-guluk in Sumenep (Madura Island) and Hidayatullah Islamic Boarding School in Balikpapan (East Kalimantan) were recognized with *Kalpataru*, the highest-order environmental award bestowed by the President of the Republic of Indonesia, in 1981 and 1984 respectively.

The *pesantren*, which has become ubiquitous in many parts of Indonesia, is a special education system, growing in line with the core teachings of Islam. Its structure consists of the *kiyai* or *pesantren* leader (cleric), *santri* (student), *pondok* (boarding), books, mosques, and other supporting facilities and infrastructure. The *kiyai* as the caregiver of the *pesantren* plays a central role in the dynamics of the *pesantren* establishments. With approximately 27,000 *pesantren* in Indonesia, and

[2] Defined by *Oxford English Dictionary* as "*a ruling on a point of Islamic law given by a recognized authority*".

[3] The movement involved the rehabilitation and reforestation of critical lands, especially those in the upstream of the river Krueng Aceh, which is an important source of clean water for the province's capital city, Banda Aceh.

around 3.6 million students (Kemenag 2019), the establishment is quite powerful and well placed as a grassroots agent of change.

With emerging environmental perspectives, some boarding schools have been transforming into *eco-pesantren*,[4] which entails several changes. To begin with, the person of the *kiyai* leads the integration of environmental protection and management to the extent of reforming the *pesantren*'s basic principles, thus expressed in the vision and mission of the new *eco-pesantren* (Mangunjaya 2014).

The *eco-pesantren* movement promotes environmental learning processes and for a while (e.g., between 2008 and 2011) some were supported by government funding. On the ground, Islam-based participatory learning modules and curricula are being developed, and environmentally friendly activities are practiced and promoted in *eco-pesantren* compounds. Activities include maintaining a healthy environment (putting into practice the *adagium* "cleanliness is part of having and practicing the faith"), recycling waste, and planting trees. With time, *eco-pesantren* networks have grown, and current leaders include Ecopesantren Daruttauhid, Bandung, Pesantren Darul Ulum Lido, Al Itifaqiyah (West Java), Annuqayah (Madura), and Nurul Haramain (Lombok).[5]

Green Mosques

The mosque has been a central and important place for Muslims throughout history—not only as a place for worship and rituals but also for sharing and internalizing knowledge and accomplishing many societal roles. Mosques represent lively grounds for discussion and debate and many have aligned themselves with environmental campaigns and movements. Recognizing that there are almost 800,000 mosques in Indonesia according to the Indonesian Mosques Council (DMI), one can appreciate how mosques hold the potential to become powerful advocates for conservation in Indonesia.

Mosque-based environmental campaigns have ramped up significantly during recent years. For instance, it was through the mosques that the Indonesian Council of Ulama *fatwas*, including those supporting wildlife protection (MUI 2014), sanitation and water management (MUI 2015), and prohibition of land and forest burning (MUI 2016) reached a large public audience. The wildlife protection *fatwa*, for instance, was deemed to have helped increase awareness among Muslims in Indonesia, providing an understanding of the Islamic perspective on protected animals; among other things, it has also helped resolve cases of human–animal conflict (Mangunjaya and Praharawati 2019).

---

[4] *Eco-pesantren* is a term denoting an Islamic boarding school with deep environmental perspectives.

[5] More than a decade ago, Nurul Haramain, under the leadership of *Tuan Guru* Hasanain Juaini, led the planting of close to a million trees and thus influenced the *pesantren* community in Lombok Island (Nugraha 2008).

One example of a green mosque comes from Masjid Baiturrahman in Yogyakarta, which has promoted seven activities: saving energy; managing waste; saving water; managing rainwater; creating orderly, neat, comfortable, and beautiful surroundings; promoting healthy and clean living; and prohibiting smoking (DMI 2015).

Another recent campaign has been to green the pilgrimage (haj), through campaigns to control the use of water, energy, and plastics. Some apps have even been developed and disseminated e.g., *Greenhajj* app or *i-Umrah*, which encourage the practice of minimizing the physical travel for pilgrimage to Mecca by doing *badal umrah* (i.e. arrangements for someone to have Umrah performed on his or her behalf by another person who is already on-site or close to Mecca). The overall aim of all these activities is to further mitigate associated greenhouse gas emissions.

Discussion

The importance of engaging proactively with the ongoing climate change and sustainability crises has become increasingly clear to many Indonesian Muslims, especially in light of the alignment between environmental conservation and many core Islamic principles. Actions such as Islamic finance mobilization, *eco-pesantren*, and green mosques are gearing up for sustainability, including safe water, sanitation, and climate change mitigation and adaptation. Out-of-the-box actions such as the purchase of community lands by religious endowments in Aceh is perhaps just one small part, but are a good contribution to inspiring countries with substantial Muslim populations.

Religion plays an important role in shaping social and economic development. In many societies, and certainly in Indonesia, religious values strongly shape political, educational, and economic institutions. Therefore, the alignment, or lack thereof, between religious principles and environmental conservation can significantly promote or hinder the process of sustainable development. A growing body of research has been devoted to the question of whether and how religious dimensions affect different facets of sustainable development (e.g., Basedau et al. 2017). The case of the growing alignment between Indonesia's Muslim community and the national and global environmental movement demonstrates the potential that communities of faith can have in promoting a sustainable future for our planet.

References

Adachi, M. 2018. "Discourses of Institutionalization of Zakat Management System in Contemporary Indonesia: Effect of the Revitalization of Islamic Economics." *International Journal of Zakat* 3, no. 1: 23–35.

ARC (Alliance of Religions and Conservation). 2009. *Many Heavens One Earth: Faith Commitments to Protect the Living Planet*, edited by Mary Colwell, Victoria Finlay, Alison Hilliard, and

Susie Weldon. Announced at Windsor, November 2–4, 2009. https://s3.amazonaws.com/berkley-center/090311ManyHeavensOneEarthFaithCommitmentsLivingPlanet.pdf.

ARC and UNDP (Alliance of Religions and Conservation and United Nations Development Programme). 2015. *Faith in the Future: The Bristol Commitments—Faith Plans for a Sustainable Future*. Bath: ARC in association with the UNDP. http://ppi.unas.ac.id/wp-content/uploads/2015/09/Faith-in-the-Future-with-cover-UN.pdf.

Arif, M. 2017. "Zakat as a Mode of Poverty Alleviation." *International Journal of Economics & Management Sciences* 6: 473. https://doi.org/10.4172/2162-6359.1000473.

Basedau, M., S. Gobien, and S. Prediger. 2017. "The Ambivalent Role of Religion for Sustainable Development: A Review of the Empirical Evidence." *GIGA Working Papers* No. 297. Hamburg: German Institute of Global and Area Studies (GIGA).

BGT (Blue and Green Tomorrow). 2015. "Environment Ministers from Over 50 Muslim Countries Agree on Declaration for Environmental Protection." October 16, 2015. Blue and Green Tomorrow website. https://blueandgreentomorrow.com/environment/environment-ministers-from-over-50-muslim-countries-agree-on-declaration-for-environmental-protection/.

Bhagwat, S., N. Dundley, and S. Harrops. 2011. "Religious Following in Biodiversity Hotspots: Challenges and Opportunities for Conservation and Development." *Conservation Letters* 4, no. 3: 234–240. https://doi.org/10.1111/j.1755-263X.2011.00169.x.

Bona, M.F. 2015. "Siaga Bumi terbentuk pada 21 September 2015." [Earth Alert Was Formed on 21 Sept. 2015]. *Berita Satu*, September 21, 2015. http://www.beritasatu.com/kesra/308682-peringati-hari-perdamaian-dunia-tokoh-lintas-agama-deklarasi-siaga-bumi.html.

Çizakça, M. 2000. *A History of Philanthropic Foundations: The Islamic World From the Seventh Century to the Present*. Istanbul: Boğaziçi University Press.

DMI (The Indonesian Mosque Council). 2015. "DMI Yogyakarta resmikan Program Baiturrahman Eco Mosque." [Yogyakarta Office of the Indonesian Council of Mosques Established the Baiturahman Eco-mosque Program]. Dewan Masjid Indonesia website. http://dmi.or.id/dmi-yogyakarta-resmikan-program-baiturrahman-eco-mosque/. Accessed December 19, 2019.

Hassan, A. 2005. "Islamic Economics and the Environment: Material Flow Analysis in Society-Nature Interrelationships." *Journal of King Abdulaziz University: Islamic Economics* 18, no. 1: 15–31.

Hutan Tersisa. n.d. "Skenario dan model konseptual untuk hutan wakaf." [Scenario and Conceptual Model for *Waqf* Forest]. Hutan-Tersisa.org website, accessed December 10, 2019. https://www.hutan-tersisa.org/p/skenario-dan-model-konseptual-hutan.html.

ISESCO (Islamic Educational Scientific and Cultural Organization). 2019. "Role of Cultural and Religious Factors in the Protection of the Environment and Sustainable Development." Eighth Islamic Conference of Environmental Ministry (ICEM), Rabat, October 2–3, 2019.

Kemenag (RI Ministry of Religious Affairs). 2019. "Statistik Pendidikan Islam dan Pesantren." [Statistics for Islamic Education and Boarding Schools] http://pendis.kemenag.go.id/file/dokumen/bukusaku1102.pdf. Accessed December 19, 2019.

Khan, M.A. 1994. *An Introduction to Islamic Economics*. International Institute of Islamic Thought (IIIT).

Lewis, M.K. 2001. "Islam and Accounting." *Accounting Forum* 25, no. 2: 103–127.

Mangunjaya, F.M. 2014. "Ekopesantren: Bagaimana Merancang Pesantren Ramah Lingkungan?." [Green Boarding School: How To Design Green Islamic Boarding Schools]. Jakarta: Yayasan Pustaka Obor.

Mangunjaya, F.M., and J.E. McKay. 2012. "Reviving an Islamic Approach for Environmental Conservation in Indonesia." *Worldviews* 16, no. 3: 286–305.

Mangunjaya, F.M., and G. Praharawati. 2019. "Fatwas on Boosting Environmental Conservation in Indonesia." *Religions* 10, no. 10: 570. https://doi.org/10.3390/rel10100570.

Mead, L. 2015. *Islamic Climate Change Declaration calls for Zero Emissions Strategy*. Accessed December 10, 2019. http://sdg.iisd.org/news/islamic-climate-change-declaration-calls-for-zero-emissions-strategy/.

Mohamed, N. 2014. "Islamic Education, Eco-ethics and Community." *Studies in the Philosophy of Education* 33: 315. https://doi.org/10.1007/s11217-013-9387-y.

Mohammad, M.T.S.H. 2011. "Towards an Islamic Social (Waqf) Bank." *International Journal of Trade, Economics and Finance* 2, no. 5: 381–386.

Mohsin, M.I.A. 2013. "Potential of Zakat in Eliminating Riba and Eradicating Poverty in Muslim Countries." *European Journal of Business and Management—Special Issue: Islamic Management and Business* 5, no. 11: 114–126.

MUI (Indonesian Council of Ulama). 2014. "Pelestarian Satwa Langka untuk Menjaga Keseimbangan Ekosistim." [Conservation of Rare Animals for the Protection of Ecosystem]. Edict no 04/2014. Jakarta: MUI.

MUI (Indonesian Council of Ulama). 2015. "Pendayagunaan Harta Zakat, Infaq, Sedekah, dan Wakaf untuk Pembangunan Sarana Air Bersih dan Sanitasi untuk Masyarakat." [Application of the Treasures of Zaqat, Infaq, Sadaqah, and Waqf to Support Development of Clean Water and Sanitation for the Communities]. Edict no 01/2015. Jakarta: MUI.

MUI (Indonesian Council of Ulama). 2016. "Hukum Pembakaran Hutan dan Lahan serta Pengendaliannya." [Edict on Burning of Forest and Land, and their Control]. Edict no 30/2016. Jakarta: MUI.

Nagaoka, S. 2014. "Resuscitation of the Antique Economic System or Novel Sustainable System? Revitalization of the Traditional Islamic Economic Institutions (Waqf and Zakat) in the Postmodern Era." *Kyoto Bulletin of Islamic Area Studies* 7, no. 1: 3–19.

NU (Nahdlathul Ulama). 2017. "Survei menunjukkan ulama lebih dipercaya daripada Presiden." [Survey Showed that Clerics are More Trusted than the President]. *NU Online*. May 13, 2007. https://www.nu.or.id/post/read/8977/hasil-survei-tunjukkan-ulama-lebih-dipercaya-daripada-presiden. Accessed December 19, 2019.

Nugraha, P. 2008. "Hasanain Juaini: Planting for a Better Future." *The Jakarta Post*. October 14, 2008. https://www.thejakartapost.com/news/2008/10/14/hasanain-juaini-planting-a-better-future.html.

Nurdeng, D. 2009. "Maintaining a Healthy Environment: An Islamic Ethical Approach." *European Journal of Social Sciences* 8, no. 4: 524–531.

Palmer, M., and V. Finlay. 2003. *Faith in Conservation: New Approaches to Religions and the Environment*. Washington, DC: World Bank.

Rahman, F. 1994. *Major Themes of the Qur'an*. Minneapolis: Bibliotheca Islamica.

Raimi, L., A. Patel, and I. Adelopo. 2014. "Corporate Social Responsibility, Waqf System and Zakat System as Faith-Based Model for Poverty Reduction." *World Journal of Entrepreneurship, Management and Sustainable Development* 10: 228–242.

SDG Philanthropy Platform. 2019. *Indonesia Green Sukuk Report 2019*. https://www.sdgphilanthropy.org/Indonesia%27s-Green-Sukuk-2019-Report. Accessed December 19, 2019.

UNFCCC (United Nations Framework Convention on Climate Change). 2015. *Islamic Declaration on Climate Change*. External Statement published August 18, 2015. https://unfccc.int/news/islamic-declaration-on-climate-change.

Chapter 21
Policy Entrepreneurship for Sustainable Development in Malaysia: A Reflection

Adnan A. Hezri

Background

Establishing a framework in the policy sphere requires the work of many individual actors, including those from the public sector. Working from within the public administration, these actors often undertake "policy work" as public sector analysts and policy researchers. Some actors also participate in public policy as "policy entrepreneurs." Unlike activists who mainly work outside of the formal government system, policy entrepreneurs work within the system or "within architecture" by acting as advisers, issue brokers, and knowledge brokers in defining policy problems and lobbying to make policymakers accept the problem and eventually take action (Mintrom and Norman 2009). Policy entrepreneurship is, therefore, a strategic act. Beeson and Stone (2013) distinguish between policy entrepreneurs and public intellectuals. A public intellectual is someone outside the formal institutions of governance who can cultivate a broad audience in championing a policy agenda. In contrast, the policy entrepreneur is a dynamic policy actor who works within the architecture of the state—sometimes behind the scenes—in pursuit of policy change. The art and craft of policy entrepreneurship include negotiation skills that allow one to function as an issue initiator and strategist to bring different people together.

This paper constitutes the author's reflection on his own two decades of experience in defining and framing policy problems to set the national agenda in the area of sustainable development in Malaysia. It is presented herein under four headings: perceived problems and prioritization; strategies and methods used for civic engagement; factors facilitating and limiting involvement; and the vision and plan for regional collaboration.

A. A. Hezri (✉)
Langkawi Development Authority, Ministry of Finance, Kedah, Malaysia

M. Indrawan et al. (eds.), *Civic Engagement in Asia*,
https://doi.org/10.1007/978-981-16-9384-7_21

Perceived Problems and Prioritization

Two trends have bedeviled the goal of environmental sustainability in Malaysia. First, conservation is hardly a top political priority in Malaysia. Although the government has formulated numerous environmental laws, enforcement is generally weak and ineffectual. Similarly, while policy statements on the environment are in abundance, government expenditure on the sector is meager. Spending on the environment and natural resource management accounts for only around 1 percent of the country's total budget. As a result, the ecological footprint of the country has increased to around 4.0 global persons per ha today from approximately 0.5 global persons per ha in the 1960s. In spatial terms, Peninsular Malaysia continues to lose its forests: forest cover fell from 9.5 million ha in 1954 down to 5.79 million ha in 2010 (Hezri 2016). A similar trend is observed in the region of Sarawak and Sabah. Second, the environmental movement in Malaysia is weakening. Elsewhere, well-organized environmental movements have proven able to exert considerable political pressure on the powers that be to strengthen environmental measures. This is not the case in Malaysia. An interview with a prominent environmentalist aptly describes the current plight in Malaysia[1]:

> *... [W]e need the base, a certain amount of broad environmental awareness and commitment. I still argue that these are still very limited. Therefore I don't even call ourselves as a movement. I see us as groups working on these issues.*
>
> *Because a movement means there's a mass of people. The first indication started with the Bukit Merah people. They are at the community level. A movement has to start at the grassroots level. We are not at the grassroots level.*
>
> *We are urbanites interacting with policy actors and all that. Of course, we try to trigger the consciousness... What we need is a crisis to trigger them [communities] and hope it catches on. Bukit Merah was a good example, but it didn't catch on because the government stopped them, Operasi Lalang.*

As indicated by these quotes, environmental activism tends to be haphazardly pursued, hence lacking the strategic push for it to become a potent enough force to mobilize stronger environmental governance (Majid-Cooke and Hezri 201). This twofold situation of low government interest and weak civil society reaction does not bode well for the quest for sustainable development in Malaysia.

Strategies and Methods for Engagement

By the late 1990s, there was an urgency to build an "inside-architecture" activism to strengthen the sustainability agenda in the country. The federal government was the primary target of this activism. From 1998 to 2011, the author was a staff member of the Institute for Environment and Development (Lestari) at the Universiti Kebangsaan

---

[1] Interview with Mr. Gurmit Singh of the Centre for Environmental Technology Development Malaysia (CETDEM) in Petaling Jaya, November 2013.

Malaysia. The institute was established in 1994 with a mission to influence policy on matters related to sustainable development. From 2011 to 2016, the author served in a renowned policy and security think tank, the Institute of Strategic and International Studies (ISIS) Malaysia, as director of environment and sustainability affairs.[2] As a novice policy entrepreneur at Lestari, the author was involved in the framing of sustainability responses at the state government level through the Sustainable Selangor and Agenda 21 initiative. The 3-year initiative led to the institutionalization of a land-use concept known as environmentally sensitive areas (or KSAS) at the state level and eventually, the national level. The author also played a significant role in the formulation of a suite of draft sustainability indicators for the Economic Planning Unit (EPU) of the Prime Minister's Department. Through these initiatives, Lestari and its researchers were influencing the definition of what sustainable development is (or is not) by incorporating ecological rationality into development discourse and planning.

Building an advocacy coalition is an important step in the art and craft of policy entrepreneurship. In 2008, the author started collaborating with one of the oldest environmental NGOs in Malaysia—The Environmental Protection Society Malaysia, or EPSM. After decades of project-based advocacy, EPSM decided to address the systemic roots of unsustainability—the institutions. The leadership in EPSM wanted to instigate a discourse on institutional change in Malaysia, which the author gladly obliged in order to assist EPSM with their high-level policy advocacy campaign called "SLIM" or Sustainable Living in Malaysia. This author was tasked with articulating the institutional challenges for sustainable development through a series of keynote addresses in various important national-level conferences.

At ISIS Malaysia, the author took a more aggressive brokering role by pursuing the attention of policymakers and the public at large simultaneously. This function constituted routine advisory and consulting activities, such as commenting on policy documents, preparing draft policy and strategies, writing speeches for ministers, and negotiating for the government in international meetings. As the project strategist at the national consultation for the Rio +20 process, the author facilitated the convening of numerous sessions with the government sector, civil society, and academia. Following this process, the author was then appointed by the EPU to draft the country's National Sustainable Development Blueprint. The draft was subsequently used by the government and the United Nations Development Programme to design a 3-year project titled Policy and Institutional Reforms for Sustainable Development in Malaysia.

While pursuing the advocacy agenda within the policy system, the author was actively involved in communicating with the public by writing op-ed commentaries and appearing on TV and radio programs to share opinions on matters related to sustainability. Finally, toward the end of his tenure at ISIS Malaysia, the author released a manifesto-type book entitled *The Sustainability Shift: Refashioning Malaysia's Future* (Hezri 2016). The book provides a clarion call for development pathways that are underpinned by the logic of sustainability. Moving forward, the

[2] The author returned to the same Institute in August 2018 as full-time staff.

book proposed that Malaysia needs to shift from framing responses through an environmental lens to a sustainable development one, and from focusing on organizations to institutions. It further proposed that rather than coming up with more aspirational goals it needs to galvanize more concrete efforts to deliver policies. Making these shifts requires nothing short of an institutional change that comprises reforms in institutional hardware, software, and "heartware."

From 2015 onwards, it became increasingly clear that there were limits to advocating policy change solely at the federal or national level. As a federation, decision-making in Malaysia on natural resources including land use, water management, and agriculture falls under the authorities of the states. Therefore "venue shopping" was necessary to create a new policy window for action at the state level. This realization resulted in a change of tack in which the author started to work sub-nationally with state-based agencies such as the Institut Darul Ridzuan (IDR) in Perak and the Penang Green Council (PGC). For IDR, the author initiated a study on "Building a Green Economy in Perak" and served as the Co-Chair of the Penang Green Agenda, an initiative sponsored by the state government of Penang through the PGC.

Factors Influencing Engagement

Persuading an audience of government leaders to accept the framing advocated by policy entrepreneurs—allowing room for open discussions and joint adaptations—can take time. In other words, there is no such thing as short-term gratification in public policy. There are numerous internal and external factors and circumstances that facilitate or limit the mainstreaming of sustainable development in Malaysia. The tempo of government in embracing environmental policies varies within different domains. It has taken the Cabinet of Ministers six years to accept the National Environmental Policy because some members viewed it as a threat to the country's industrialization drive. Similarly, the endorsement of the national solid waste management bill has taken almost ten years (Hezri 2016). This has to do with the fact that some states viewed the federalization of waste management as an infringement on their powers. In contrast, the government took only 100 days to formulate and endorse the National Green Technology Policy, as they viewed green technology as a profitable new economic sector. As for sustainable development, the government's official position is that Malaysia has integrated the concept into its planning since the 1990s by subscribing to the principle of "balanced development." In practice, however, the words may be found in the right places as policy statements, but the actions are not (Hezri and Nordin 2006). This tactical stance to do less, in effect, limits the absorptive capacity of key agencies, such as the EPU, to embrace ambitious targets and goals related to environmental policies.

The effectiveness of a policy entrepreneur depends on the political resources that s/he commands. In the context of sustainability in Malaysia, these resources are somewhat scarce. True to the paradox of the connectivity era, the management silo is not just a problem in the government circle, but also among other key development stakeholders. The rather limited space for critical feedback on government policies has led to acquiescence or self-censorship on the part of the scientific fraternity, academia, and think tanks on matters related to land-use decision-making, for example, to the detriment of the environment and the populace.

Vision for Collaboration

The global endorsement of the 17 Sustainable Development Goals (SDGs) with its 169 targets has wedged open the most significant policy window to date for sustainability. Underpinned by the principles of "leaving no one behind" and "indivisibility of the goals," the SDGs also beg a different form of policy entrepreneurship on the part of actors working to influence public policy. Recognizing this, in 2015 a group of policy brokers established an informal network called the Malaysian Civil Society Alliance for Sustainable Development Goals (henceforth CSO Alliance). The CSO Alliance currently comprises over 40 civil society groups from the human rights, environment, social work, social business, and gender domains. The author is a co-founder of the CSO Alliance together with a renowned sociologist and public intellectual, Dr. Denison Jayasooria. Together, the Co-Chairs orchestrate the activities of network members who mostly share beliefs about the problem of unsustainability in Malaysia and collectively seek out solutions.

Apart from engaging with the government in the preparation of official SDG documents, such as the 2017 Voluntary National Review report to the High-Level Political Platform, the CSO Alliance aims to function as a "think-and-do tank" on matters related to SDGs in the forthcoming years. With the membership of prominent civil society leaders, it is envisaged that the Alliance will grow to be a political coalition with more significant weight in influencing policy change toward sustainable development. In 2019, the Speaker of the Malaysian Parliament gave agreement for the Alliance to form one of the All-Party Parliamentary Groups (APPGs), a new structure of informal cross-party groups mandated to provide feedback on various issues to the Parliamentarians. This recognition represents an institutional breakthrough on the part of the Alliance and also its Co-Chairs as policy entrepreneurs aiming to create a new economic model based on the principles of sustainable development.

Conclusions

Despite government pronouncements on the importance of sustainable development in Malaysia's policy direction, the unpalatable reality points to its lackadaisical implementation. With low awareness among Malaysians of the importance of environmental sustainability, this policy domain has remained a marginal concern among policymakers until today. Confronting this double whammy situation requires entrepreneurial policy change strategies. The reflection offered in this chapter highlights the choice of strategies of an advocate of sustainable development policy working from within the architecture of the state. The policy window for change is currently larger than it has been in the last few decades with the victory of the Pakatan Harapan government at the federal level on May 9, 2018. Moving forward will require all policy entrepreneurs and members of the epistemic community to join forces in building a stronger coalition that pushes for a truly balanced development throughout the country.

References

Beeson, M., and D. Stone. 2013. "The Changing Fortunes of a Policy Entrepreneur: The Case of Ross Garnaut." *Australian Journal of Political Science* 48, no. 1: 1–14.

Hezri, A.A. 2016. *The Sustainability Shift: Refashioning Malaysia's Future.* Penang: Areca Books.

Hezri, A.A., and M. Nordin. 2006. "Towards Sustainable Development? The Evolution of Environmental Policy in Malaysia." *Natural Resources Forum* 30, no. 1: 37–50.

Majid-Cooke, F., and A.A. Hezri. 2017. "Malaysia: Structure and Agency of the Environmental Movement." In *Routledge Handbook of the Environment in Southeast Asia,* edited by P. Hirsch: 400–417. Routledge: Abingdon.

Mintrom, M., and P. Norman. 2009. "Policy Entrepreneurship and Policy Change." *Policy Studies Journal* 37, no. 4: 649–667.

Chapter 22
Step By Step From Cambodia Toward ASEAN

Heng Monychenda

Introduction

It is thrilling to recall that 35 years ago, specifically in July 1987, the Republic of Indonesia's Minister of Foreign Affairs, Mochtar Kusumaatmadja, Chairman of the ASEAN Standing Committee, successfully initiated the "cocktail party" that later developed into the three Jakarta Informal Meetings. These initiatives worked toward solving the issue of Vietnam's invasion of Cambodia, which some observers called the Third Indochina War, from 1978 to 1991. These efforts eventually led to the signing of the Paris Peace Agreements on October 23rd, 1991. Allow me to say "Thank You, Indonesia" for such a wonderful gift of peace in action to Cambodia, to Indochina, and ASEAN as a whole.

I would like to emphasize in this paper three venerable concepts that a prominent monk of Cambodia introduced to us from the time of the Jakarta Informal Meetings, in which he played a key role. These concepts, consistently articulated by His Holiness Maha Ghosananda, are still followed by many people, even though he passed away in 2006. First is the idea he often expressed that peace is possible and that it is possible to make peace our common goal. Second, peace needs time and careful preparations to be realized and lived by all people from all walks of life and by all religious communities. And third, a responsible and mindful community working for peace and following the Middle Path will contribute to a peaceful nation, which will, in turn, contribute to a peaceful world. If I, we, and future generations, our children and their children, could act while holding these concepts dear, we could live in a peaceful, sustainable ASEAN in times to come.

H. Monychenda (✉)
Buddhism for Development, Sanger, Cambodia

M. Indrawan et al. (eds.), *Civic Engagement in Asia*,
https://doi.org/10.1007/978-981-16-9384-7_22

My Remembrance of Maha Ghosananda's Work

In JIM I (1988) His Holiness Maha Ghosananda led contingents of Buddhist monks to the peace negotiations in Jakarta, Indonesia. These negotiations had brought together the four main warring factions in Cambodia, including the faction supported by Vietnam. Maha Ghosananda's fifth contingent was unique, however. Adopting a strictly neutral stance, it was conceived as an army of peace, whose ammunition would be "bullets of loving-kindness; an army absolutely without guns or partisan politics, an army of reconciliation with the courage to use nonviolence to solve problems, an army dedicated wholly to peace and the end of suffering" (Santidhammo 2007, 41). With this contingent, he proposed compromise and reminded Cambodian leaders that, *"Peace is our common goal. Peace is possible!"* (Santidhammo 2007, 41).

I was fortunate to be able to come to Indonesia with Maha Ghosananda for JIM II in 1989. As the Secretary of the Cambodian Mission for Peace, he was the one responsible for these contingents whose members were willing to face the armies of guns with love and compassion and to spread the message of reconciliation, forgiveness, and peace to end the suffering of Cambodian people.

The Cambodian Mission for Peace was only to be the first and less well-known realization of the army of peace that Maha Ghosananda envisioned. These diplomatic peace contingents later transformed into the *Dhammayatra*, the Dhamma Walk, a walk for reconciliation, peace, and development that continues until today.[1] The first Dhamma Walk held in 1992 began with training in nonviolence and group decision-making at a Cambodian refugee camp inside Thailand. From there, little more than a hundred refugees walked through contested Khmer Rouge territory to Phnom Penh. However, hundreds more joined as the walk made its way through local villages. It was not unusual to see tearful reunions as walkers encountered long-lost relatives in villages passed along the way.

The first walk demonstrated that it could be an effective way to overcome fear and trauma and to enable people to work together for peace. And the *Dhammayatras* have continued to play this role every subsequent year. In March 2006, we commemorated the 10th Anniversary of Maha Ghosananda's death with the 27th *Dhammayatra* to promote the message of a peaceful heart with the Five Precepts he had taught us:

1) Peace is our common goal. Peace is possible!
2) We need to remember that our temple is always with us. Our temple is within us.

[1] The Dhamma Walk initiated by Maha Ghosananda was crucial in inspiring the organization of similar Dhamma Walks in Thailand, which began with an international Peace Walk in 1995 in commemoration of the 50th anniversary of the Hiroshima bombing. Thai participants in that walk subsequently organized nine similar walks around Songkhla Lake, and Dhamma Walks continue to this day in Chaiyaphum Province, led by the Thai monk, Phra Paisal Visalo. Participants themselves have referred to these walks variously as Dhamma Walks, Dhamma Marches, or Peace Walks. (Theodore Mayer, personal communication) The Khmer version is sometimes used in English language sources: Dhammayietra.

3) All religions walk together seeking peace of the heart one step at a time, one step at a time.
4) Hatred never ceases by hatred in this world but by love alone is healed. This is the ancient and eternal law.
5) The suffering of Cambodia has been deep. From this suffering comes great compassion. Great compassion makes a peaceful heart. A peaceful heart makes a peaceful person. A peaceful person makes a peaceful family. A peaceful family makes a peaceful community. A peaceful community makes a peaceful nation. A peaceful nation makes a peaceful world."

We still bring up this central message of His Holiness Maha Ghosananda again and again, every year, to remind ourselves not to forget to apply it and to spread it to others. Now, on the ground, the *Dhammayatra* remains influential in governing the behavior of the people in Cambodia, especially the students of primary and high schools, to be aware of HIV/AIDS transmission, to prevent the use of drugs, and to join together to protect and preserve our environment.

Is Peace Possible? Peace as a Common ASEAN Goal: Issues and Hopes

As a result of the Indonesian initiative of 1987, a politically and culturally fragmented Cambodia facing conflict and violence[2] moved toward internal peace in a very basic sense, namely "the state of the absence of war."[3] The government of Cambodia is very proud of this definition, while many do not fully agree with it.[4] It is what Martin Luther King Jr. called "negative peace" (King Jr. 1963). Many believe that there should be rule of law, human rights and responsibilities, human dignity, democracy, freedom, and importantly sustainable development, to make a living peace meaningful in the twenty-first century. Still, peace as the absence of war and indirectly of violence has become an important topic of discussion in many countries of Southeast Asia. It allows a space where "positive peace" can take root and grow.

The struggle between the government and the nongovernment formations in Cambodia has certainly polarized the two groups (and in a way complicated civic engagement). The situation has resulted in a competitive spirit, bringing democracy and peace into direct conflict.[5] This has "naturally" led to the conflict between

[2] So severe was the fragmentation that I used to call it the "Cambodia of Pieces".

[3] From Dr. Mahathir Mohammad, who has often used this phrase. For a very recent example, see Associated Press, "Progress Realizable in Absence of War," March 10, 2019.

[4] Albert Einstein, for example, signed a document, "Appeal to the Peoples of the World," unanimously approved by a Conference on World Government at Rollins College in Florida, held in 1946. The document argues that peace is not only the absence of war, but the presence of justice, law, and order. See Nathan and Norden (1968, 371).

[5] Unhealthy competition takes on life when the effort to do good following one's own way becomes in the end the only right way.

government and nongovernment organizations. A new term appeared in Cambodian social media early in 2019, which will likely lead to another great debate: kleptocracy, i.e., government by those who seek status and personal gain at the expense of the governed.[6] This is another area of possible conflict.

What we are afraid of is that our peace of mind in Cambodia will be broken into many pieces and thus plant seeds of dissension. It may spread out to ASEAN, the greater nation I have gradually fallen in love with. Although ASEAN is still at the stage of the ASEAN Economic Community, due to financial and economic reasons, ASEAN will survive and live in peace and prosperity only by nurturing a common vision, identity, and community…only in peace.

Looking at ASEAN, peace has been threatened in many of its member countries due to internal and external factors. The durability of peace has been questioned in the minds of people: how long will the absence of war in the Korean Peninsula be feasible, and how big of an impact will it have on peace in ASEAN? How far will the negotiations on the flashpoints in the South China Sea hold together before the absence of war breaks down? How much more time and valuable resources must be spent and how many more lives will have to be lost before the war on drugs is won? Last but not least, to what extent can we stop provoking war and violence in the name of religion, which causes pain and suffering across many communities?

In spite of all these realities, we still believe that peace is possible. It is our sacred common goal. We need a skillful team with not only the best strategy to gain and grow peace but also with the best strategy to prevent the loss of peace. This team must elicit the participation of the people in realizing the ASEAN vision, ASEAN identity, and most importantly, the ASEAN community. So far, these things have been dictated by politicians and business corporations. But peace is only possible when it is our common goal—not simply the goal of politicians.

Can All Religions Walk Together Seeking Peace of the Heart, One Step at a Time?

I do not like to limit this walk and this work only to religions, for this approach is too institutional. Hundreds of ethnic groups with different languages and beliefs must be mobilized in seeking and promoting the peace of the heart. One of the main problems for people who identify with one particular belief system, however, is the feeling that other belief systems are alien, bad, not good enough, or imperfect, and the attempt to enculturate such feelings in their children and the next generations. Then they bond this feeling to nationalism, patriotism, or even territorialism, and colonialism as well. Buddhism, my religion, is not an exception.

There is a great diversity of beliefs and practices among the indigenous peoples of Cambodia, who have often been profound models for me of peace and compassion of the heart. We need to actively engage indigenous communities. We need to invite

[6] "Kleptocracy" is a new term in Cambodia, though it was coined in 1819.

indigenous people to dialog and work together with the main religious groups to promote peace in ASEAN—one step at a time toward peace.

There are many differences in terms of our national or ethnic cultures and in terms of our cultural or political institutions. But our basic behaviors related to eating, sleeping, being afraid, being caring or kind, sexual reproduction, and so on are all the same. Our religions have taught us to be better persons in terms of the culture of love, empathetic compassion, sympathetic joy, and justice. These cultures aim at both physical and mental peace in terms of our everyday coexistence. Unfortunately, sometimes we are limited by the view that "only my culture can bring peace to my community and the world." Therefore, it is time to find the compatibilities between belief systems, particularly as they relate to peace and compassion so that these cultures can interrelate in various specific areas without problems.

Peace is the common goal of all religions—Islam, Christianity, Buddhism, Shintoism, and Hinduism, among others. A common vision of ASEAN should include peace as the core—*not* economics and profit-making. Yet we must remember that understanding peace may differ among the various followers and politicians and these understandings may not be compatible and may result in conflicts. We have to reach for and expand the core beliefs of religions surrounding peace. Our scholars have found and agreed upon several core social teachings of each religion that are not only compatible but also lead to peaceful coexistence. However, the core social teachings leading to peaceful coexistence are rarely taught in schools to promote harmony, togetherness, and a sense of common humanity among the children. Fear takes charge here. Each religious group is afraid that the teaching of other religious principles may lead to loss of one's own religious identity. And, when religious identity is linked with ethnic or national identity, such religion-based courses may lead to threats to ethnic or national identity. It is a big obstacle to achieving peace in ASEAN.

If we are open, we will find that the most compatible core social teachings of Buddhism and Islam are moderation and justice i.e., the *Majjhimāpaṭipadā,* The Middle Path, and the *Ummatan Wasaṭan*, the Community of the Middle Way. Such compatibilities lead to understanding our differences, dialoguing to arrive at some commonalities and peacefully coexisting. It is the way of the Buddha and the Prophet.

Buddhism

Lord Buddha (Tathagata) who was staying near Varanasi (India) at a deer park at Isipatana addressed a group of five ascetics.

There are these two extremes that are not to be indulged in by one who has gone forth. Which two? That which is devoted to sensual pleasure regarding sensual objects: base, vulgar, common, ignoble, unprofitable; and that which is devoted to self-affliction: painful, ignoble, unprofitable. Avoiding both of these extremes, the middle way realized by the Tathagata—producing vision, producing knowledge—leads to calm, to direct knowledge, to self-awakening, to Unbinding.

And what is the middle way …? Precisely this Noble Eightfold Path: right view, right resolve, right speech, right action, right livelihood, right effort, right mindfulness, right concentration. (Dhammacakkapavattana Sutta).

Islam

> "And thus have we willed you to be a community of the middle way, so that (with your lives) you might bear witness to the truth before all mankind" (Qur'an, Al-Baqara 2:143). "Ummatan Wasatan is a community that keeps an equitable balance between extremes and is realistic in its appreciation of man's nature and possibilities, rejecting both licentiousness and exaggerated asceticism (Asad 1984, 30)."[7]

Christianity

There is no exact equivalent to the concept of the Middle Way in the Bible, and there are many different explanations of how the Middle Way could be applied in Christianity. However, thirty years ago a paragraph from the Bible made me understand the Middle Way of Christianity from my point of view as a Buddhist monk. I was impressed with: "He humbled you, causing you to be hungry, yet he fed you with manna that neither you nor your ancestors had known, to teach you that human beings do not live by food alone — instead human beings are to live by every word that proceeds from the mouth of the Lord." (Deut. 8:3)

When the most compatible core of religions has been defined and if it is well accepted by those concerned, all other teachings would be less difficult to employ as a means to end the suffering of mankind. It may be difficult, but it is possible: by taking one step at a time all religions can walk together seeking and promoting peace.

Civic Engagement: Can a Peaceful Community Make a Just and Sustainable Nation? The Case of Buddhism for Development

All governments are always pushing the people, directly or indirectly, to engage in the development of the country. A Cambodian Civic Education curriculum has been produced for high school students in the hope that they would positively change their knowledge, skills, and attitudes toward valuing other human beings, the culture of peace, and the engagement in social development (economic, cultural, and political).

[7] From Muhammad Asad, *The Message of the Qur'an*, p. 30, cited by Yusuf Imtiyaz in "Dialogue Between Islam and Buddhism" (2009, 384).

However, this curriculum has not included or focused on STEM subjects, which makes it less interesting for the children and the teachers. The tendency to push an educational model has been minimal on the part of the government. However, there are several civil society organizations, including Buddhism for Development (BFD), that have tried to promote some subjects, or some aspects of core subjects, particularly those addressing people's participation in the democratization of the country and promotion of a culture of peace. Nongovernmental groups' involvement in the drive for a democratization of the government very often leads to suspicion. This is due to the many sensitive subjects that are addressed, namely human rights, corruption, accountability, distribution of power, and so on. All these concerns are seen as something that might lead to weakening the functioning of the government.

Buddhism For Development (BFD) is a non-governmental, not-for-profit, non-party, and non-partisan community development organization founded on 1 May 1990 at Prasat Serey Monastery, in Refugee Camp Site 2, by the author, who was a Buddhist monk with the ordained name of Indapanno Bhikkhu, along with the monks I was leading at the time. It was based on the understanding of the value of Buddhism as a comprehensive tool for reconciliation, peace restoration, national reintegration, economic development, social development, mental development, wisdom development, human rights recovery, the building of democracy, and sustaining the balance between individuals, society, and nature.

At the center of the BFD logo is the Dhammacaka Wheel, symbolizing movement toward the Middle Way and the Eightfold Path of Life. Inscribed within the wheel are the words, "*Caratha Bhikkhave Carikan*," referring to the Buddha's instruction: "Go your ways, oh monks, for the benefit and happiness of many, out of compassion for the world, for the good, benefit, and happiness of gods and men ..... There are beings with only a little dust in their eyes who will be lost of *Shanti* (peace) unless they hear the *Dhamma*."[8] Bodhi leaves stand for the environment and its nurturing protection, the lotus flower below for the support and participation of target groups, and the halo for loving-kindness and unlimited compassion. Khmer and Roman letters together represent indigenous culture and international cooperation.

The logo of BFD is in fact about active civic engagement; that is the message of the Compassionate Buddha. The following activities are some examples of BFD's engagement.

The Indigenous Culture

As a young monk with very little experience, I faced difficulties when I started engaging with local people after returning from the displaced person's camp near the Thai-Cambodian border in 1992. It was even more difficult to get their active support due to my status as a "newcomer" in my natal village (though I was a monk and the son of a village leader).

I recalled a Khmer proverb that says: "Love the daughter to take care of the mother, love the mother to protect the daughter." This led to the first project of BFD, the *Kindergarten of Reconciliation*. In this project, children from all the conflicting political factions in Cambodia were selected if they were from poor families, had disabled parents, or were orphans. They were then provided with a meaningful education. The children were happy, the parents were happy. Then we continued with other projects to take care of older parents and the less fortunate families. Today, our scholarship program continues to operate in six northwestern provinces of Cambodia.

The insight I drew from the Khmer proverb has helped me to take steps toward achieving two key points in my philosophy for civic engagement: reconciliation of the nation and its peaceful development.

The Middle Path

The success of BFD in the development of Cambodia, unfortunately, has come along with some challenges, particularly from officers who live with the following extreme attitude: "*Those who are not my friends are (or maybe are) my enemies.*" They do not believe in the Middle Path or neutrality in daily life. In 1997 when I returned from Boston during the fall break, I encountered a direct and challenging question. A high officer and a longtime cadre of the party in power approached me just one day after my arrival at my office in Battambang, Cambodia. He asked me: "Since you left the monkhood, what stand are you going to take?" I found the question odd. My quick answer was: "I was a monk for a long time. When I went out to get food

[8] *Vinaya Pitaka, Mahavagga, First Khandaka* 11.2. https://realtruthlife.blogspot.com/2011/03/mahavagga-first-khandaka.html#.XRM_Lh7VIWo.

in the morning, I never asked whether the owners of the food came from this or that political party. I am grateful to them all. Buddhist monks are not allowed to take sides in party political affairs (royal affairs in the Buddha's time). Although I am no longer a monk, the idea of not being involved in partisan politics is still very strong in my mind and conviction.[9] Also, BFD is an NGO and an NGO has its status as a non-partisan social entity by law." The man left me with a few short words: "Thank you, I got what I wanted."

My story is just to illustrate that the Middle Path is not easy to implement. Most people find it difficult to understand. I think it is one of the critical characters of civic engagement in Cambodia—the use of the Middle Path, or the Ummatan Wasatan, in a world of extreme polarization.

Basic Humanity

Under the Communist Khmer Rouge, I and others were warned to remember four dictates: not to see, not to hear, not to know, and not to talk. Not to heed this warning meant risking death. Although the situation has improved during the last ten years, these imperative keywords still produce a negative impact today. BFD promotes five positive keywords as cross-cutting themes to reverse the impacts of the old dictates throughout all activities: to observe, to listen, to learn, to complain, and to demand. This reversal works as a bottom-up approach. Through these new keywords, people learn that the eyes, the ears, the mouth, and the brain belong to all people who are willing to use them for the happiness of their families and communities, and to challenge the exclusive rights of a few to own those wonderful organs. We cannot be any better than other creatures if our basic humanity is taken away!

Volunteerism

Life moves forward. It can be a development or a deterioration. The next generation must be the subject for discussion on various topics so that we can make sure that our descendants are ready to meet unexpected situations. Over the last 10 years, BFD has provided opportunities for approximately 1,500 university and senior high school students to voluntarily work in the subnational and local administration. These opportunities enabled them to understand more about real-life governments and their constituencies. They taught them to give away what they have to fulfill the needs of others and to prepare themselves to continue their mission for a just and sustainable Cambodia.

---

[9] My conviction derives in part from my father's warning from even before I became a monk not to involve myself in party politics.

Between 1999 and 2011, when some parts of Cambodia were in the fragile state of peace characterized only by the absence of war, this kind of volunteerism helped to harmonize people who had come from different ideological backgrounds: socialism, royalism, Khmer Rouge ideology, and liberalism. The volunteers helped people to reintegrate their emotional and physical lives in 110 communes of seven provinces in the north and west of Cambodia. Each village was encouraged to elect one volunteer who would be called a Peace and Development Volunteer (PDV) to be trained by BFD. Those who became PDVs would go out from one house to another to discuss the issue of "trust" among their communities. As agents of change, the PDVs also acted as facilitators for alternative dispute resolution in the village. At the commune level volunteers from the villages were included in the formation of the Commune Committee to Prevent and Manage Human Rights Violations (CPMHRV), together with the members of Commune Councils and other respected community leaders. Networks to Prevent Human Rights Violations were also established at the district level with the recognition of the district authorities.

BDF also promoted volunteerism in another area, access to health, which has always been a problem in Cambodia. Village Health Volunteers were established through BFD's health program in 1995 to disseminate health information and provide basic advice on health care issues. In 1997, the model was studied by the Ministry of Health along with other available models and has become part of a national policy on health care volunteers. Today 47 trained volunteers—all people living with HIV/AIDS—are working to serve 2700 others also living with HIV/AIDS in 11 districts of Battambang province.

My belief about volunteerism is that every one of us can be an agent of change in our community, regardless of where we live. The impact is great, and the expense is less, because the ones who encounter the same problems together can sort the problems out among themselves. They own the responsibility; thus they engage with their heart to sustain the goodness they have done.

The Malaria Model

I got malaria several times during the period when the Khmer Rouge ruled the country, between 1975 and 1978. I always wondered how fast the disease spreads in our body, starting with only one bite of an agent-mosquito. I learned later in the displaced people's camp that the *Plasmodium falciparum* multiplies itself eight times over. This drove me to the idea that in development we often find we cannot multiply a successful activity with sufficient speed; but if others see the goodness of the activity for their communities, they can multiply that activity at a far faster speed than I could alone.

Most of the projects initiated by BFD take into consideration the possibilities for multiplication in the future in related areas, according to the needs of each situation. For example, starting from a single FM radio program, "Buddhism for Khmer Society," a live program on various engagements that Buddhists can take on to improve quality of life in Battambang, there are now hundreds of programs on FM radio conducted by Buddhist monks and lay Buddhists in all 25 provinces. BFD started the program in 2007, then followed up with training and capacity building for monks in several provinces for three years. Ten years later hundreds of programs had multiplied from this beginning: "Buddhism and Peace," "Buddhism and Harmony," "Buddhism and Khmer Literature," "Buddhism and Quality of Life," and so on. In Battambang province alone there are 35 civic engagement programs on a dozen FM stations. They are all financially self-sustaining due to contributions from the listeners (Buddhists) and have even been able to raise money and materials to support poor children and families and to provide emergency assistance.

The Malaria Model needs well-qualified projects, support by the target groups, and social-oriented benefits. Multiplication is not replication; rather, projects need to be adopted and adapted by communities according to their socioeconomic circumstances and culture.

Conclusion

I believe my contribution through Buddhism for Development to my country does have a positive impact on ASEAN. The impact may be small, but together with other civilians of Cambodia, I believe we can make an observable change toward a more just and more sustainable Cambodia. If we can do this, then we certainly are contributing to the progress of ASEAN. My dream for the progress of ASEAN would be that it be a borderless region, made up of peaceful communities, characterized by only a small income gap between the nations and harmony among its diverse ethnicities and religions.

One cannot say that civic engagement for a just and sustainable ASEAN (and Cambodia) is developing smoothly now; it may be bumpier than ever. The power of social media, especially Facebook, allows, on the one hand, a better understanding of the work of government and its constituencies. But on the other hand, the dark side of governments has been exposed to the public, with thousands of comments and shares criticizing and blaming the government for many reasons. Governments have used national security, the color revolution, the cost of gaining and maintaining peace, and other issues to reduce the engagement of civil society actors. The exercise of freedom of speech in the public and in social media leads to the question of "freedom after speech" in many countries in ASEAN (what happens, in other words, to those who speak out). Social media literature is no longer so much the problem, but ethics and morality have become the main worries both for government and nongovernment groups.

As a bottom-up, non-violent, and antiauthoritarian approach that relies on consensus decision-making, civic engagement is truly time-consuming. This may be one reason it is not put into action that often. Yet the creation of a civic engagement movement is a worthy goal and should be the next step toward integrating ASEAN at a higher level, based on grassroots initiatives. I am now in the early stages of establishing a Middle Way Movement within ASEAN, by which I hope to bring together groups in our countries that follow the Ummatan Wasatan of Islam, or the Middle Way of Buddhism, Christianity, or Hinduism. The rise of extremism, both left and right, which usually involves real gun violence, encourages me to set up this civic movement. It may take time, but I have already taken the first step, using social media. As Maha Ghosananda said, one step at a time, one step at a time, we can achieve peace within the hearts of the ASEAN population. "Put down the Gun and take up the *Dhamma* (the Middle Way)" is the motto I have believed in for the past 30 years.

"*Natthi santi param sukham*"—There is no greater happiness than peace!

References

Asad, M. 1984. *The Message of the Qur'an.* Gibraltar: Dar al-Andalus.

Associated Press. 2019. "Progress Realizable in Absence of War." *Manila Times,* March 10, 2019. https://www.manilatimes.net/progress-realizable-in-absence-of-war/523187/.

Deuteronomy 8:3 *International Standard Version* of the *Bible.* Bible Hub website: https://biblehub.com/isv/deuteronomy/8.htm.

"Dhammacakkapavattana Sutta, Setting the Wheel of Dhamma in Motion." In *Samyutta Nikaya,* LVI.11. https://www.buddhismtoday.com/english/texts/samyutta/sn56-11.html.

Imtiyaz, Y. 2009." Dialogue Between Islam and Buddhism through the Concepts *Ummatan Wasatan* (The Middle Nation) and *Majjhima-Patipada* (The Middle Way)." *Islamic Studies* 48, no. 3.

King Jr., M.L. 1963. "Letter from a Birmingham Jail." Originally published as "The Negro is Your Brother." In *The Atlantic Monthly,* 212, no. 2: 78–88. August 1963.

Nathan, O., and H. Norden. 1968. *Einstein on Peace.* New York: Schocken Books.

Qur'an, Al-Baqara 2:143. Quoted in "Give Da'wah, Awaken the Hearts." Da'wah Skill website, September 19, 2016. http://www.dawahskills.com/abcs-of-dawah/give-dawah-awaken-hearts/.

Santidhammo Bhikkhu. 2007. *Maha Ghosananda: The Buddha of the Battlefield.* Woodinville, Washington: Atammayatarama Buddhist Monastery.

Vinaya Pitaka, Mahavagga, First Khandaka (The Admission to the Order of Bhikkhus) 11.2.

Chapter 23
Civic Engagement for a Just and Sustainable ASEAN

Erna Witoelar

Highlights

- Civic engagement is progressing in the world.
- Seeing the bigger picture and embracing a larger mission helps fight intolerance, build goodwill, and leave no one behind.
- Sustainable Development Goals (SDGs) are local, and therefore, each locality must address its own context.
- As both the processes and products of the SDGs are rich in knowledge, all concerned parties need to understand the essence of the SDGs.

Introduction

One can feel the excitement of the privilege—witnessing, supporting, and appreciating the progressive endeavor of Asia's public intellectuals and activists. These are people with such rich backgrounds who are so committed to civic engagement, especially at the community and local levels, and who at the same time have managed to influence national and global policies. Here, I wish merely to add some perspective to what Asia's public intellectuals and activists are already doing and experiencing.

I want to start with civic engagement at the global level. One can see how organizations, ideas, commitments, and idealism already exist and are integrated into various platforms. Through what **process** do we relate to such platforms? I will then concentrate on the Sustainable Development Goals (SDGs) themselves. How do the SDGs, as a **product**, influence what we are doing, and the work of transformation more broadly? Finally, I will discuss the **challenges** and **opportunities** of civic engagement in the regional context, focusing on partnerships and synergies.

E. Witoelar (✉)
KEHATI (Indonesian Biodiversity Conservation Trust Fund), South Jakarta, Indonesia
e-mail: erna@witoelar.com

M. Indrawan et al. (eds.), *Civic Engagement in Asia*,
https://doi.org/10.1007/978-981-16-9384-7_23

The Process

We all know that civic engagement on the global level has existed for many, many decades. My life's work concentrated on the environment and sustainable development, but I know that many Asian public intellectuals and activists have been very involved in various dialogues around no less pertinent issues, such as gender equality, religious harmony, and protecting the rights of indigenous peoples. These initiatives are now already making a significant difference.

In the arena of the environment, the United Nations Summit on the Human Environment in Stockholm in 1972 was a starting point, but it was government-oriented. It was not until the Rio Earth Summit in 1992 that we began to see more civic engagement. However, even while civil society representatives were there, we were not effective participants. We could not even enter the summit. We were just standing in the lobby, and we were lucky if we could even get our messages to one or two of the delegates, to try to influence them.

After the Rio summit, though, there was slowly more and more civic engagement at the global level, so that by the second Rio, Rio+20, and during the meetings afterward, I was very much surprised and overwhelmed by the level of engagement. We could see a variety of civic actors, even young adults, who before had only been in the lobby, now presenting at the summit, inside the UN building, and talking to the UN heads of states.

The Millennium Development Goals (MDGs) were one step in reaching our goals, but civic engagement with the MDGs was very, very slow. Our governments in Asia were also very slow. The MDG process was considered a project of the UN, and the UN published direct country reports on progress toward the goals. This created a feeling that decision-making was being taken from us. All of this increased the motivation for stronger civic engagement with the SDGs.

SDGs

One important milestone in the journey was 1983, when the sustainable development concept was first conceived and people moved away from viewing the environment in isolation to analyzing and linking the economic and social dimensions of environmental problems. Although this holistic approach is not fully successful yet, hopefully now with the SDGs, we can do better.

What has changed, today? For one, women's issues have come to the fore. I remember, in all those conferences, stronger and stronger movements grew up around women and sustainable development, and around including women's issues in the MDGs. Now, through the process of setting the SDGs, we see gender mentioned in almost all the goals. So now, gender equality is not just one (separate) goal, and it is not just a problem for women: it is everybody's challenge. Whatever goal you are working on, whether you are working on job creation or any other issue, gender is an important and integral part of the picture.

The SDGs are for everybody. The SDGs are very strong in promoting human rights, with particular emphasis on women, youth, and the elderly. The difficult challenge, however, is governance. With the MDGs, which were driven by national governments and international experts, we could still apply sectoral approaches, working within each ministry (health, women, and so forth). Not surprisingly, though, private sector actions were not properly registered in national planning.

The SDGs, in contrast, are more integrative. For instance, water and sanitation concerns have to be addressed in all relevant sectors, including industry and labor markets. With such multi-sectoral requirements, iterative multiple stakeholder processes are urgently needed. In this way, the SDGs are more inclusive. In Indonesia though, many non-state actors proceeded without consulting the government. The good thing is that, as envisioned by our former president "Gus Dur" (K.H. Abdurrahman Wahid), the space for civil society was provided, with the public servants taking on more of a facilitation role. It remains true, though, that the national government also needs to coordinate and provide clearinghouse functions through Indonesia's multi-layered provincial and local governments. A new (reformed) governance, one that includes innovative and new partnerships, is therefore needed as the "glue" to hold the various actors together on the path toward achieving the SDGs.

Challenges and Opportunities

Learning from the MDGs, we see that there is a wide-open space. Many people were involved in reaching MDG targets, and there is leadership in the organizations. This is good news. Yet with the wider and more diverse issues and concerns of the SDGs, as well as the increased challenge of governance, partnership, and regional leadership, there must be transformative learning. How do we catalyze this transformation and make it useful for the people we have been working for?

I am going to elaborate on this a little bit because I lead an organization called Filantropi Indonesia. We work with multiple types of philanthropy: family, corporate, and religious-based philanthropy, as well as philanthropy for research, for gender, and a host of other issues. Things are moving very fast in the field of philanthropy. It is already on people's minds, and because people have started working on all issues related to the SDGs, we are promoting deeper engagement. In short, we encourage people to reach the SDGs from wherever they come from.

One of the very exciting initiatives we have started at Filantropi Indonesia is to help move the religious-based organizations from charity to development and, from there, to form partnerships with other religions. This is good because all religions are teaching us to be generous. Many Asian cultures are noted for being generous. We are automatically giving every year, every month, to people close to us. So we are all philanthropists at different scales. Philanthropy organizations expand this to a much bigger scale. But we, individually, are giving. Now, in Indonesia, we are also giving in the religious context of Islam, and also within the Christian, Buddhist, and Hindu contexts. Our organization facilitates the movement of this giving to allow it to do something more. Because the SDGs are so wide and so noble in their aim to reach everybody, to leave no one behind, they have to be supported by different means of implementation and different sources of financing.

Indonesia has been quite active in philanthropy and in expanding on the idea of working with friends. Due to the presence of fundamentalism and other problems in our religions, there are more and more efforts to build solidarity within and across religions. Doing this has required plenty of dialogues and building partnerships based on those dialogues, which began with inter-religious dialogues on the environment. These efforts aim to move people away from conflicts by giving them a bigger mission to handle together. In this way, conflicts may be eased or forgotten. This is one realm that I would like to encourage regional actors to study and engage in their work.

The processes to achieve the global pledges of the SDGs and other recent agreements are quite inclusive; they are built on stronger ownership and stronger commitments. And so, I am quite excited about the idea of everybody doing something to achieve the SDGs on whatever level they can.

In Indonesia, a presidential decree passed in 2016[1] strongly recognizes the role of non-state organizations and civic engagement in the realization of the SDGs. This decree provides a legal basis at the national level for our work; we are working on achieving this also at the local level. The latest decentralization policy,[2] however, has presented a challenge in that the subnational governing authorities in most development sectors are assigned to provinces (not cities and districts). In other words, the subnational action plan for sustainable development (derived from the national action plan) reaches the provincial level only. It may be assumed that provinces will support cities and districts to develop their action plans. But this, of course, does not always happen, and the process is often subject to the relations and dynamics between each province and city or district. In response, the Ministry of Planning and the provincial planning boards have devised fiscal incentives to increase the motivation of cities and districts to engage with the SDGs. Since it will take a long time to reach the local governments, Filantropi Indonesia has opted to bypass the local bureaucracies by working with the *associations* as well as *select* entities of cities and district governments (i.e., those with progressive leaders) to better educate them.

Almost all of the SDGs are local goals. Whatever we want to achieve, if we do not work at the local level, we will not meet the SDGs. So it has to be at the local level. Yet unfortunately, whatever commitments, whatever tools, whatever resources or technologies are involved, they do not often come down to the local level due to lack of information, access, and so on. That is why we need bridging from activists, who know the local level very well, who know the resources that exist at the national and regional levels, and who can provide assurance to the communities who will need to use those resources.

[1] Republic of Indonesia's Presidential Regulation no 59/ 2016 regarding Implementation of Sustainable Development Goals.

[2] Republic of Indonesia's Law 23/ 2014 regarding Local Government.

The SDGs came out of the sustainable development concept, which, at the beginning, was promoted as including People (social), Planet (the environment), and Profit (the economy). Profit was later changed to Prosperity, and by the time the SDGs were launched, the fourth and fifth elements of Peace and Partnership had been added. This completed the concept of sustainable development and is the basis of the present SDGs. This was a transformation from the singular goal-by-goal approach of the MDGs to an interdependent approach with the SDGs.

This interdependence is exciting. There are many things to study and to learn about linkages as well as partnerships. The governments should be pushed to produce proper reporting, rather than simply providing data, and to integrate the SDGs with local and national development. We advocate starting with the priorities of the local governments, because, for example, the priorities for the Archipelago of Maluku are different from the priorities of the province of Central Kalimantan. Priorities should not be decided at the national level. And we are happy that slowly some of the national officials are becoming quite understanding of this concept, including the Ministry of Home Affairs, which formerly often did things in a top-down way. In the past, ministers who had a military background led this ministry; but now, officials are gaining a better understanding.

It is important to deepen engagement. With whatever one is doing, there are many roads, many potential partners. Many potential transformations need to be developed and studied. In doing this, I believe we must always start from the local level, which is where the strength of the people on the ground is rooted.

I would like to finish by discussing the challenges of arriving at a just and sustainable region, such as ASEAN. Changes are inevitable. So, transformation is not just a keyword for one's program, it is a keyword for anyone wishing to achieve changes. Leadership styles are also important; it is not easy to pull together diverse interests and actors, in diverse modes and environments, in a way that can build durable networks and partnerships. We need leadership to facilitate that task as well. Some people call it "bridging leadership" or "facilitating leadership." It is a far more challenging kind of leadership than it is when we work to engage people who are already committed, for with them we do not even need to talk much about the complicated elements of the SDGs. Yet it is a deeply rewarding experience when one finds that through bridging leadership one can support the development of strategic partnerships. For we see that effective alliances between diverse actors are able to bring forth and facilitate the full dynamics and virtues of civic engagement. One beautiful pattern that has emerged from my latest engagement in advocating and supporting philanthropy through Filantropi Indonesia[3] is that when all is said and done, equitable and dignified multi-stakeholder approaches increase sustainability.

[3] Perkumpulan Filantropi Indonesia (PFI), the legal entity of Filantropi Indonesia, together with sustainability-sensitized associations such as Indonesia Business Council for Sustainable Development (IBSCD), Indonesia Global Compact Network (IGCN), and Indonesian Chamber of Commerce (Kadin), also founded the Philanthropy and Business Platform for SDGs (FBI4SDGs) in 2018. The Platform's membership presently comprises 15 associations representing about 500 Indonesian foundations and corporations. Encouragingly, for each network, multi-stakeholder approaches are in progress and being further developed.

I would also like to link more across the region because this is an exciting time for the regional approach. We do not work in silos anymore; we try to work together. Within ASEAN, we have been a family for some time, and much as within a family, there are sometimes ups and downs. But the principles of the ASEAN Charter are relevant to whatever context we are working in, whether it is for religious solidarity, stronger gender equality, a better understanding of sustainable development, addressing climate change, or other goals. I think with this experiential capital, we can do many, many things, whether it is across the region or just between two or three countries. We have already done this, so we do not need to reinvent everything to expand our work.

Looking ahead, unlike the UN Framework Convention on Climate Change, especially the Paris Agreement, which is a binding agreement among governments, the SDGs are not. They are a set of concepts that if applied, can bring us closer to what we need to reach. So try comprehending them, because both the processes and products of the SDGs are rich in knowledge, in commitment, and in materials for future study and transformative learning. They offer a very clear and suitable way to move the people of the region toward long-lasting partnerships.

Chapter 24
Reflections on Civic Engagement and Key Issues Raised in This Volume

Chheang Vannarith, Maung Maung Yin, and Theodore Mayer

Introduction

All three of the authors of this chapter participated in the *Civic Engagement for a Just and Sustainable ASEAN: Our Stories and Practices* workshop that took place in Yogyakarta from August 11–15, 2017, and from which this volume emerged. We were among those invited to share our own stories and practices there. We want to first express our gratitude for the tremendous behind-the-scenes work undertaken by Michiko Yoshida of Chulalongkorn University, and by the staff members of the Indonesian Consortium for Religious Studies, who together made that historic workshop possible. Their joint efforts made for an occasion that was full of hope and the possibility of deeper understanding and further collaboration. It was a bringing together of extraordinarily diverse perspectives from Asian activists and public intellectuals who were commendably willing to share their first-hand experiences. Many of the participants were institution builders and long-term champions of civic engagement. We remember too that in considering what steps to take next, it was Mochamad Indrawan, one of the original members of the civic engagement initiative, who proposed an edited volume that would extend our own discussions to a larger public and to policymakers. We are grateful to him for steadfastly seeing this work through to completion.

C. Vannarith (✉)
ISEAS-Yusof Ishak Institute, Singapore, Singapore

M. M. Yin
Professor of Christian Social Ethics and Peace Studies Founding Director of Peace Studies Center, Myanmar Institute of Theology, Yangon, Myanmar

T. Mayer
Institute for Transformative Learning of the International Network of Engaged Buddhists, Thailand
e-mail: director@inebinstitute.org

M. Indrawan et al. (eds.), *Civic Engagement in Asia*,
https://doi.org/10.1007/978-981-16-9384-7_24

At this moment in history, people around the world share the concern and even the anxiety that neither national governments nor the institutions of global governance seem to be especially resilient; in the face of COVID-19, we may be forgiven for wondering if they are even viable as the key agents and problem-solvers on behalf of the public good. The pandemic has made the most basic competence of some national governments an honorable and life-saving quality, regardless of political stance (Durning 2020).[1] Yet the difficulties go much further. We seem to be entering a new era of crises driven by rising protectionism, ethnoreligious extremism, climate change and environmental degradation, and geopolitical rivalry between major powers. And the core economic structures that drive income and wealth inequality prioritize high rates of extraction (both labor and raw materials), and leave large gaps in realizing development—all these features remain firmly in place, led by the corporate and the financial sector. They also have slowed the response to the climate crisis and the shift to renewable sources and sustainable levels of energy use, thus ensuring that climate breakdown will continue to be a core threat to the planet and to humanity. We see that existing neoliberal institutions have not led to a just and sustainable society, and many institutions still tend to work in silos. Within Southeast Asia, the credibility of governments remains low and has even declined due to a lack of popular participation and, in some cases, to conditions that have made such participation virtually impossible.

But all is not lost. This volume shows us that in response to governance and legitimacy crises, in response to the irresponsibility of many corporate boards, civic engagement that welcomes deeply transformative learning can change our mindsets and re-design our national and global institutions. We need to better coordinate and strengthen such civic engagements, and eventually re-connect regional institutions under a new and humane order. ASEAN, a key regional body of Southeast Asia, has taken progressive steps in the past, such as declaring Southeast Asia a nuclear-weapons-free zone.[2] Were ASEAN to now put social and planetary sustainability, and the welfare of its people, at the heart of its regional cooperation, learning, and exchange, it would be setting a rare and valuable precedent for all such regional bodies around the world. Yet the weak responses of ASEAN as a body to massive human rights violations perpetrated by some member states in recent years raise serious questions as to whether ASEAN can play such a historic role. Perhaps the governmental and intergovernmental failures we have seen only highlight the importance of citizen action and people-to-people diplomacy. How, though, can self-organized grassroots action ultimately link up with and influence governmental and intergovernmental policies? This volume offers a number of examples and shows us that there is hope for expanding civil society's participation in policy dialog, design, and implementation. We see from the concrete examples collected here that ordinary people

[1] Alan Durning, 2020, "The Plague Brought the Renaissance. What Could COVID-19 Bring? Three Hypotheses on Post-Pandemic Life." Sightline Institute website, May 8, 2020. https://www.sightline.org/2020/05/08/the-plague-brought-the-renaissance-what-could-covid-19-bring/.

[2] The Southeast Asian Nuclear-Weapon-Free Zone Treaty (Treaty of Bangkok) entered into force on March 28, 1997 (NTI 2022). https://www.nti.org/education-center/treaties-and-regimes/southeast-asian-nuclear-weapon-free-zone-seanwfz-treaty-bangkok-treaty/.

can be the drivers of genuine development; they *can* develop innovative and creative solutions through social and ecological enterprises. Similarly, those within the sphere of government *can* innovate through policy entrepreneurship and by doing so assist governments to work honestly with communities toward justice and sustainability.

We can identify in this volume some of the key steps needed to address the challenges we face at many levels. These include civil society actors engaging policymakers and stakeholders in an inclusive manner; inspiring and working with young people to shape a development trajectory that is more equitable, peaceful, sustainable, and resilient; creating participatory models of development that put people at the center and that build on the possibilities for multi-stakeholder relationships (including state-market-society partnerships); taking a middle way between different religions and civilizations and promoting cultural tolerance and the appreciation of diversity; valorizing the "heartware"[3] of connection to a tradition of practices and values and to a particular geographical setting; documenting and sharing knowledge, including local as well as endogenous knowledge; and fostering solidarity and cooperation among civil society groups that go beyond the borders of the nation-state.

The chapters in this volume also help us to see the sources of power that allow civil society actors to engage effectively with their constituent communities as well as with policymakers. Unmistakable here is the raw passion that drives individuals and small groups to try to set things right whatever it takes, but also their willingness to learn from their experiments. Present also is the humility required for such learning and for compromise, both essential to creating durable networks and alliances. Through all this we see the slow accumulation of hard-earned knowledge and the sprouting of innovative and inspiring ideas. Other sources of power for civil society are the power of telling stories and listening to people's concerns. This volume also extends that opportunity to a wider circle of those who would listen.

Some Lenses to Employ for Reflecting on the Contributions of This Volume

One thing this volume offers in abundance is a dynamic view of "the grassroots." If ordinary people's participation in decision-making and in solving problems is important as a foundation for democracy, then the chapters here provide us with a multidimensional perspective on what "the grassroots" actually means in Asia and how ordinary people engage as members of civil society. In fact it is difficult to find a chapter in this volume that does not in some way talk about the grassroots.

Indigenous communities figure prominently in a number of chapters as key protagonists of grassroots civic engagement in Asia (Chaps. 1–3, and 7). Members of these communities appear in the volume as defenders of forest lands and of their right to be involved in policymaking regarding land, forests, and forest products. They also provide examples of the role of "heartware" in sustaining social practices that in turn

[3] See Dicky Sofjan, Chapter 3.

sustain the waters and other lifegiving features of the environment. They include the conservers of traditions of woven cloth that rely on local dyes, symbolic systems, and production practices (Chap. 7).

Indigenous defenders of the land and of the environment are also closely connected to a theme that runs through many chapters, namely sustainable farming. Sustainable farming in turn connects directly with a very broad grassroots category, which is consumers. For we are all consumers, and we all want to be able to eat food that is healthy, nutritious, and unadulterated by toxins. Here direct farmer-consumer linkages seem particularly hopeful. Grassroots citizen scientists, who have popularized methods for measuring toxicity more generally and who advocate for protective policies, are also present in chapters on Thailand and on the Fukushima triple disaster of 2011. The chapters that deal with broad popular participation in urban decision-making and planning (Chaps. 8 and 10) would seem to be on the other end of the spectrum from the indigenous grassroots. Yet the themes of information sharing, constructing knowledge through experience, and building alliances of people to engage with policymakers are present there too. Reading this broad range of stories allows us to imagine a democratic future society shaped by the needs and the ideas of the grassroots, in which culturally diverse but harmonious cities would evolve in respectful relationship with sustainable forest communities and healthy rural townships.

One dimension that cannot be missed in this volume is the vital role that women now play in civic engagement in Asia. Chapter 15 on "Female-Driven Climate And Environmental Action: Champions From Pakistan," by Areej Riaz and Mairi Dupar, is the one chapter that focuses explicitly on the significant role that women have already played in the area of sustainability. This chapter suggests further that removing remaining obstacles to women's full participation will be crucial to achieving a livable and stable planetary environment. Chapter 23 by Erna Witoelor echoes this claim, based on decades of work in the movement for sustainability internationally. Yet a quick look at the list of authors for this volume will also reveal that no fewer than 16 of the 31 authors of these chapters are women. Of those, at least eight are founders, co-founders, or directors of an independent and well-functioning civic engagement program or organization. Another female author is responsible for instituting key policy shifts and awareness programs in response to the threat of poisonous jellyfish in Thailand. Like all the chapters in this volume, these women's accounts speak with the authority that only comes from bold action, persistent inquiry, the building of alliances with like-minded friends, and openness to a transformation of perspectives brought about by their own work and experience in a given area of civic engagement.

Nearly all of the chapters in this volume discuss efforts to influence policy and policymaking. Two of the chapters are by policymakers themselves, Chapter 16, by Alex Rumaseb, and Chapter 21, by Hezri Adnan. Rumaseb's chapter is unique in the volume as it is essentially the reflections of a high-level planner with many years' experience in Indonesia's Papua and West Papua provinces on the complex conditions that have led to armed conflict and to a lack of sustainable development in that geographical region. We are invited to consider the impacts of the Indonesian government's policy of encouraging mass transmigration of ethnic Javanese into Papua and

West Papua provinces along with an emphasis on extraction of local resources, while also taking into account the lack of unity of the indigenous population as they have faced development designed by others. Hezri Adnan's chapter discusses his two decades worth of experience as a policy entrepreneur focusing on sustainability in Malaysia. A policy entrepreneur is, in his view, "a dynamic policy actor who works within the architecture of the state – sometimes behind the scenes – in pursuit of policy change." Both chapters are rather rare windows into the personal reflections of policy actors committed to sustainability.

The volume as a whole presents diverse examples of interventions that have aimed at or resulted in policy (re)design and implementation. A prime example is the Livelihood Sovereignty Alliance (LISO), a coalition of three Vietnamese civil society organizations working with indigenous ethnic minority communities, who together played a key role in the passing of Vietnam's 2017 Law on Forestry (Chap. 1). Another example is the dynamic intervention on smart and sustainable urban development by the Kota Kita Foundation in Indonesia (Chap. 10). Its mapping and problem-solving strategies have been adopted as part of the national government's urban planning policy. Other countries, such as Mongolia, have also learned from the experience of the foundation.

Serious interventions aimed at public awareness, policy advocacy, and capacity building have also had positive impacts on the issue of human trafficking. The Labor Rights Promotion Network Foundation (LPN) of Thailand has rescued about 3,000 victims of human trafficking in the fishery sector in Thailand and across Southeast Asia (Chap. 14). Raising public awareness of human trafficking has affected consumer behavior and drawn public support to fight against the brutal practices involved. Public opinion, in turn, has shaped government policy and intervention. Now the Thai government and local and international organizations are cooperating with LPN in collectively addressing the issue.

The civic engagement initiatives of other contributors often include efforts to influence or change policies and legal frameworks as one part of a broader program that focuses more on building new forms of knowledge and thus changing practices at the local level. Nestor Castro's piece (Chap. 8) shows how a policy-framing tool such as the SDGs could be made a part of a transformative learning process in which local communities reflected on the SDGs, identified the ones they felt were critical to their lives, and then chose moral qualities embedded in their own traditions that could encourage work to realize the goal. The latter is an example of a conceptual tool at the heart of Dicky Sofjan's Chapter 3, a tool that in some ways permeates the volume. It involves distinguishing the hardware, software, and "heartware" elements of sustainability, where heartware refers to the dimension of faith, deeply held values, and spiritual practices that support the concepts and practices of working for balance between human communities, nature, and that which is seen as sacred. Chapter 20, on Islam and environmentalism in Indonesia, similarly focuses on the role of heartware in promoting sustainability.

Finally, every chapter reveals some dimension of the transformative learning process that has taken place within and through the various forms of civic engagement that the volume highlights. This appears in different guises in each chapter. Many authors reflect on their own learning process. A number of the chapters describe knowledge construction leading to changes in perspective and practice as explicit goals of their broader activities. In a few cases, such work even led to the construction of learning modules to be included in school curricula. Theodore Mayer's chapter on an English and leadership program he designed for young adults in Asia proposes reorienting university-level classrooms towards a wider transformative aim, and identifies integrity, connection, and play as key values and components within that shift. Several chapters similarly focus on cultivating leadership qualities, whether it be among youth in Laos or farmers in Vietnam.

Issues and Approaches—Some Examples in Greater Depth

This volume considers sustainable development challenges in many areas. These include disaster management, human trafficking, urban planning and development, indigenous people's rights, sustainable forestry and sustainable agriculture (organic farming), social enterprises (textiles and agri-business), social and religious harmony, capacity building and networking support for civil society groups, knowledge creation and sharing, and leadership training.

Each chapter presents the mission, struggles, and passions of groups and individuals as they work to bridge urgent gaps and fulfill specific societal needs. Leadership skills, personal dedication, and commitment determined how issues were selected and dealt with in pursuit of the sustainability mission. Issues taken up within a given chapter are often complex and tend to reinforce or provide counterpoint or context to the issues explored in others. We explore some of those interconnections and overlaps here.

In the spirit of capacity building, the Sustainable Agriculture Foundation (Thailand-SAFT) has promoted public awareness of the importance of food safety and sustainable agriculture in Thailand. SAFT engages and connects farmers and consumers, as well as local and national governments. Its focus on food and food production has successfully affected social behavior around the consumption of organic food and has contributed to the safe food movement in Thailand.

SAFT was particularly concerned with how the monopolization of food supply chains has impacted food security. People need to be aware of what they are eating, and may even need to be educated about food production and food security, since many of us know little about what we eat. This issue goes far beyond Thailand, of course. One example would be that in Myanmar and much of India, where many people appreciate spicy food, local media sources give attention from time to time to stories about how something as simple as chili powder may be adulterated with brick powder, talcum powder, salt, starch, wood husk, or artificial dyes meant to dye fabric, not food.[4]

SAFT addresses the dire need for civic engagement to develop knowledge on food production, supply, and security, especially among the wider public. It has shown that situation analysis and participatory action research undertaken together with farmers, joint seed conservation and research, developing farmers' markets, and fostering entrepreneurship through farmer schools, are all useful for food security while supporting the autonomy and decision-making of farmers. SAFT's creation of social spaces for farmers, urban citizens, and those interested in agriculture and farming to hold serious discussions about farming practices and sustainability has led to transformative learning on all sides as well as to changes in concrete practices. These changes benefit both farmers' income and agricultural sustainability that protects consumers from vulnerability. In this way, SAFT provides a key lesson that learning through example as well as joint research and discussion empowers farmers to become producers, entrepreneurs, and owners of the market.

In the present era of climate breakdown, such excellent work at the national level will most likely be increasingly impacted by regional and planetary issues as well. The interplay between rice agriculture and climate change is one example. While rice is the staple food for vast populations throughout much of Asia, research on climate change and greenhouse gas emissions estimates that a considerable amount—11% of global anthropogenic methane emissions—is produced from rice agriculture (IPCC 2013). While this is not good news for Asians, scientists are working on developing and adapting new species of paddy seeds as well as new planting approaches to meet the challenge of reducing agriculture's contribution to greenhouse gas emissions. Perhaps this is the kind of issue in which participatory research by farmers, measurement techniques devised by citizen scientists, and lab research by agricultural technicians would find occasion to come together. In the high biodiversity context of Indonesia, people are beginning to realize that local foods—such as breadfruit, yam, and taro, and for West Papua in particular, sago—can play a crucial role in combatting climate change. These local foods help reduce dependence on rice paddy monoculture, which has a significantly higher carbon footprint (Mochamad Indrawan, personal communication).

The Participatory Development Training Center (PADETC) in Laos focuses much of its efforts on capacity building and leadership training, but in their case the efforts are now focused on youth, who make up a large portion of the Lao population. Youth became the focus of PADETC's efforts in 2009, and by the year 2018 some 331 rural youth between the ages of 16–26 had graduated from its 4-month intensive training course. They received training in a variety of areas including leadership, team-building, organic farming, village participatory research, gender analysis, and environmental management. We see that the national context is of great importance, as development schemes in Laos have largely been top-down affairs that do not

[4] This article at NDTV's website would be one example of many from India available in English (NDTV 2018). https://food.ndtv.com/food-drinks/is-your-red-chilli-powder-pure-4-simple-tests-to-find-out-1787847

reach or involve the participation of remote areas. This is a limiting reality; yet that limitation begins to be challenged when even government officials notice that grassroots initiatives *can* reach and involve those remote areas of society. PADETC's work and that of other CSOs was given greater legitimacy with a new government legal framework first laid out in 2009, that allowed CSOs to function and to participate more fully in the public sphere. Nevertheless, issues of the legitimacy of grassroots initiatives continue—with the still unresolved disappearance in 2012 of the founder of PADETC, Sombath Somphone, and with the attitudes of some in the Lao government who may still view CSOs as disrupting social harmony.

The multi-stakeholder approach to civic engagement can be observed in some detail in many of the chapters of this volume. We think it will be especially productive if readers can look carefully at the interactions between different stakeholders in each chapter to see exactly who they are and what their interactions reveal. This volume is rich in examples, and summarizing them all is beyond the scope of this chapter. We do urge the reader to mine these examples for what they yield in lessons, insight, and understanding. Below we provide a brief example of what we mean.

A key goal of the Livelihood Sovereignty Alliance (LISO) of Vietnam is to make it possible for indigenous forest communities to manage forest resources, have a secure legal relationship to the land, and continue their way of life—all based on their own customary practices and values. In Vietnam, as in many developing countries, this is difficult because land is treated as belonging to the population as a whole; the state then administers the land on behalf of the people. This conflicts with the customary communal land tenure of indigenous forest-dwelling groups. LISO manages to work within these national restrictions by working first to legitimize customary communal tenure with *local* rather than with national government officials. The process has worked as follows. After the land in such indigenous communities is allocated according to customary practices and beliefs, the communities then draw these up as land-use regulations. The community's own land-use regulations thus produced are then brought to local government authorities responsible for land management to receive their official recognition. Once these customary practices have been in place for some time and have proven reliably successful in responding to communities' needs, while managing local ecologies safe from threats of external land-grabbing, they become models of effective local land governance. LISO then presents them as such to the national government. In this way, customary practices had a powerful influence on the reformed Vietnam Law on Forestry (2017), which recognized community ownership of spirit forests, traditional watershed forests, and traditional production forests.

By following the trail of interactions between diverse stakeholders, we can see how a strategy for influencing policy developed under very specific conditions. We can also witness a deeper truth revealed throughout this volume—that whenever a local group manages to successfully resolve an issue, their specific solutions become models worthy of study, emulation, and application by others. Every chapter holds some brilliance of this kind, whether it be Sekar Kawung's multipronged approach to supporting indigenous woven fabrics through exhibitions, school curricula, various forms of media, and direct connection with the online fashion world; or a careful assessment of what kind of partnerships and conditions led in Lombok to the most enduring and successful installation of household biogas plants; or the Samdhana Institute's long and varied experience working to realize their vision of "a region where natural, cultural, and spiritual diversity is valued and environmental conflicts are resolved peacefully, with justice and equity for all parties;" or Hiroko Aihara's bringing together of all those who have suffered as a result of nuclear waste, accident, or bombing into a Global Hibakusha Network; or the emergence of Green Mosques in Indonesia; or the multilayered approach of the Asian Public Intellectuals (API) Regional Project on "Community-Based Initiatives toward Ecological Balance" carried out over many years in five very different communities across Asia.

This brilliance does not deny the reality of failures, negative outcomes, and unintended consequences. In fact, the authors should be commended for boldly sharing some of the difficulties and unpleasant realities they encountered. Recipients of sincere help on some occasions become passive or take what they can without further participation. Social and political conditions may be extremely challenging for a variety of reasons, whether from corruption, a fear of ceding power prerogatives, or simple prejudice. The five communities studied in the API project were not all successful examples of traditional values enabling sustainability in the face of large-scale development or extractive enterprises. Many worthwhile projects stumble for lack of sustainable funding structures. Some projects, without proper and inclusive design, may lead to local conflicts and even community disintegration. And trust can be eroded if a project does not produce expected outcomes. Yet without these challenges and failures, would we see the extent and depth of learning that these chapters reveal? As Surichai Wun'gaeo's Foreword suggests, transformative learning may not come without some serious challenge or existential dilemma. Perhaps that is another form of richness we can look forward to in these uncertain times.

Scope for a Regional Approach—Some Considerations and Questions

In this volume we see individuals and groups in ASEAN and other regions of Asia coming to terms with three issues: *civic engagement, justice, and sustainability.* Although there is no easy consensus on the definition of these terms, for us *civic engagement* refers to the participation of civil society and grassroots organizations in policy design and implementation *as well as* in the building and testing of innovative models for resolving a myriad of issues at the local, regional, and/or global level. These models may or may not lead to new perspectives and policy changes at the level of formal structures; and we acknowledge that it is a genuine victory when they do. Yet the important thing is that someone would have had the courage and the energy to try something new and to share what they learned in doing so more broadly. This is one reason that we see transformative learning as so essential to our understanding of civic engagement. *Justice* refers to an ASEAN or Asian community that respects human rights and human dignity, and *sustainability* refers to communities and regions that care about the environment, ecosystems, and future generations. We see this volume as clearly a step forward for collaboration and sharing of lessons between very differently positioned leaders on these themes in Asia. It seems appropriate to ask what could now further strengthen this process of sharing more widely and building further possibilities for collaboration?

One response is simply to take heart upon reading the successful initiatives of others, then (with or without fanfare) apply their models and strategies elsewhere. How could the participatory research of farmers or the approach of key farmers be applied elsewhere? How could the clarity and participation engendered by Samdhana Institute's mapping be applied to domains and issues more broadly in Asia? Who among us could not learn from the savvy use of media by Sekar Kawung or the Labor Protection Network, and consider how they could be put to work elsewhere? SAFT, SENS, Kota Kita, and others represented in this volume all make the exemplary work of grassroots leaders and thinkers part of their own leadership training curricula—whether through field study, site visits, or extended interviews. In this way, the intellectual and moral wealth of a given society is made directly available to young NGO, urban, or farmer activists. How could we expand this face-to-face learning process across borders and regions, either physically or digitally?

We should not forget that parallel to the publication of this volume, a core of individuals has gone on to build on the original Yogyakarta workshop, as Professor Surichai outlined in the Foreword. Subsequent events and conferences have relied on the readiness of university officials and the resources of willing international donors and foundations to create innovative platforms for bringing together young leaders, grassroots activists, university professionals, religious thinkers, and government officials to engage in nontraditional forms of encounter and exchange. Efforts such as the Bangkok Forum or the Civic Engagement 4.0 International Forum in Solo truly represent steps taken by very diverse actors to meet each other outside of traditional silos so as to solve what can appear to be overwhelmingly broad problems of sustainability. How can we build on this process?

Professor Surichai's call to meet each other primarily as human beings, human beings who face common challenges, with profound willingness to respect, listen, and learn about the very different conditions of life that have shaped us and the aspirations that guide us—this would seem to be an important ethical call at the heart level. Southeast Asia is noted for its incredible diversity of religious traditions, landscapes, and social histories, a diversity still dwarfed by that of Asia as a whole. But as a starting place, perhaps Southeast Asian cultural mores of politeness and mutual respect in civil discourse could be allowed to come to the fore in our debates, so that none are disrespected for disagreeing or for not knowing. We say this fully aware that those mores have broken down into unimaginable violence on numerous occasions in Southeast Asian history, including recently. This awareness raises the question as to the "why" of this breakdown and how it can be averted. It also reminds us of the urgency of working towards the "negative peace" that Cambodians treasured, which is the state of being free of war and violence (Chap. 22). It seems to some of us that the great suffering of the COVID-19 pandemic has also created the mood for a shift in public attention to the importance of basic human values and solidarity. In this shift, a reliance on proven scientific approaches and the need to speak honestly about what will lead us to essential human securities has become paramount. The sheer fact of the pandemic, however, is unlikely to raise to a higher level the serious questions about the fall-out from our profit-driven economies and power politics once the pandemic is over. How could those sincerely engaged in the quest for a sustainable future ensure that these questions are faced calmly and openly in the coming period? And how could they use the platforms of their civic engagement work to raise these deeper questions more broadly?

Due to the informal nature of much civic engagement work, finding forms of sustainable funding remains an important question for nearly all of us. Surely the approach of social entrepreneurship that we see examples of in this volume is one answer. On the other hand, we need to remember something that we are already very good at. That is the importance of investment that is not only of money—however necessary that is—but also of time and attention in people and in building trust. This includes in leaders or in any person, from whatever position they hold in society, who is capable of doing some genuine good for the people and for our societies within real planetary boundaries.

Concluding Remarks

Civic engagement, indispensable for good governance and sustainable development, is one of the means to make this world a place where all people can enjoy increased human security and environmental sustainability. Let us continue to spread our stories and practices to help break the chains of ignorance and irresponsibility. Let us be prepared for the risks and challenges facing us, and be brave enough to innovate in our region and beyond. Let us take courage from the stories assembled here of our fellow thinkers, designers, researchers, and practitioners, whichever walk of life they may happen to come from, and especially those from the dynamic and energized grassroots. Whatever credibility, power to transform, or merit we have together accumulated, whether through learning, experience, hard work, or the building of trusting relationships, let us generously share these with an ever larger community around us. Let us fulfill this essential duty we have learned through civic engagement of working to hear, honor, and learn from every voice. And let us pass on to our children and grandchildren, the future generations, the potential for progress and the rich inheritance of insight, learning, and human solidarity that civic engagement makes possible, so that theirs will be a benign future that nourishes all forms of life.

References

Durning, A. 2020. "The Plague Brought the Renaissance. What Could COVID-19 Bring? Three Hypotheses on Post-Pandemic Life." Sightline Institute website, May 8, 2020. https://www.sightline.org/2020/05/08/the-plague-brought-the-renaissance-what-could-covid-19-bring/

IPCC. 2013. *Climate Change 2013: The Physical Science Basis. Contribution of Working Group I to the Fifth Assessment Report of the Intergovernmental Panel on Climate Change.* Cambridge University Press, Cambridge, UK and New York, NY. https://www.ipcc.ch/site/assets/uploads/2018/02/WG1AR5_all_final.pdf.

NDTV. 2018. "Is Your Red Chilli Powder Pure? 4 Simple Tests to Find Out." NDTV Food Desk, at the NDTV website, March 19, 2018. https://food.ndtv.com/food-drinks/is-your-red-chilli-powder-pure-4-simple-tests-to-find-out-1787847.

NTI. 2022. "Bangkok Treaty - Southeast Asian Nuclear-Weapon-Free-Zone (SEANWFZ) Treaty." NTI website. https://www.nti.org/education-center/treaties-and-regimes/southeast-asian-nuclear-weapon-free-zone-seanwfz-treaty-bangkok-treaty/

Index

M. Indrawan et al. (eds.), *Civic Engagement in Asia*,
https://doi.org/10.1007/978-981-16-9384-7

The manufacturer's authorised representative in the EU is Springer Nature Customer Service Centre GmbH, Europaplatz 3, 69115 Heidelberg, Germany. If you have any concerns regarding our products, please contact ProductSafety@springernature.com

Printed and bound by CPI Group (UK) Ltd, Croydon, CR0 4YY
30/09/2026
02225242-0008